# TILLAGE

A practical guide to the latest tillage methods, conservation planning, crop residue management, and solutions to soil problems

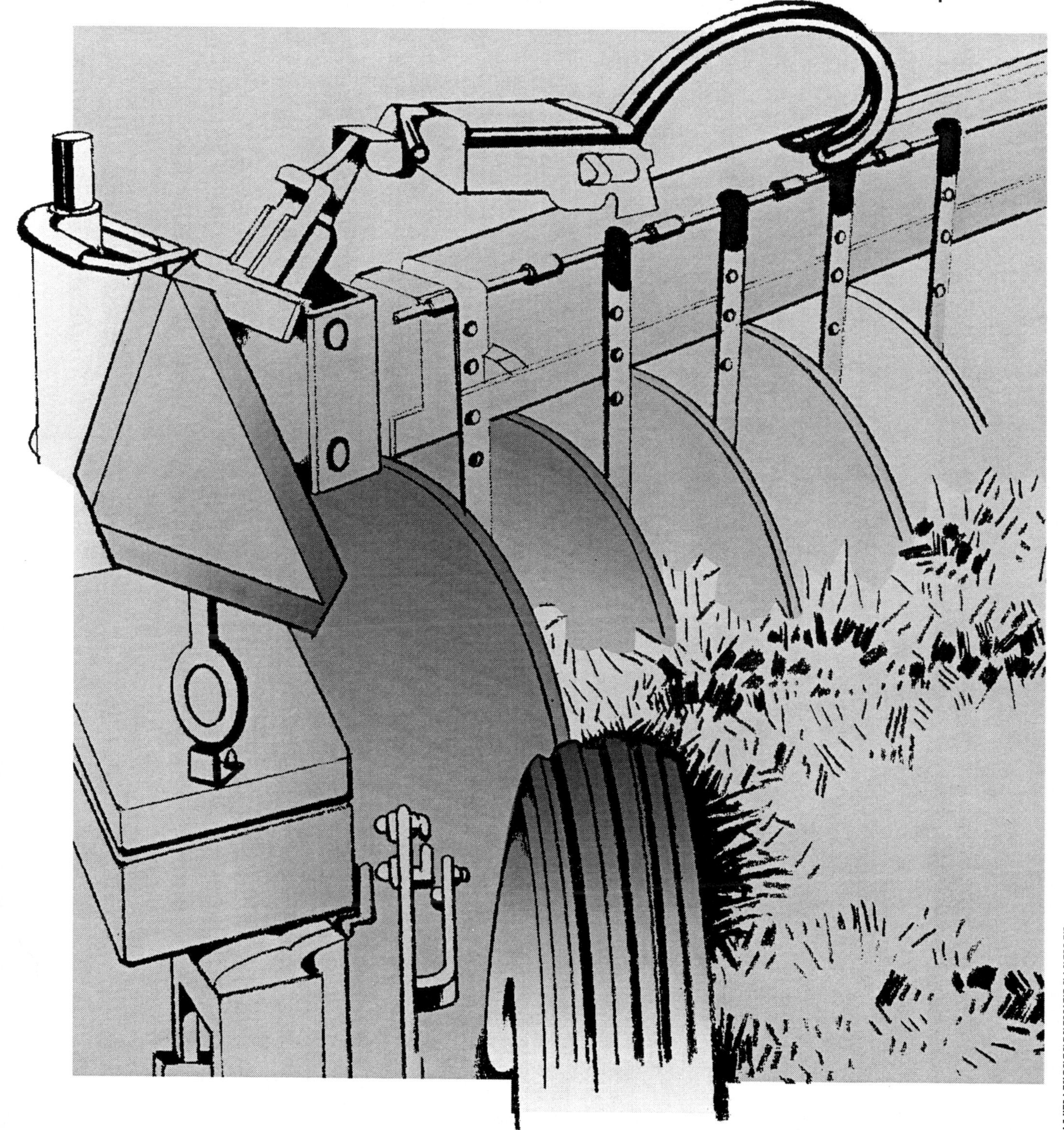

FUNDAMENTALS OF MACHINE OPERATION

**PUBLISHER**
DEERE & COMPANY
JOHN DEERE PUBLISHING
one John Deere Place
Moline, IL 61265
http://www.johndeere.com/publications
1–800–522–7448

Fundamentals of Service (FOS) is a series of manuals created by Deere & Company. Each book in the series is conceived, researched, outlined, edited, and published by Deere & Company, John Deere Publishing. Authors are selected to provide a basic technical manuscript that could be edited and rewritten by staff editors.

HOW TO USE THE MANUAL: This manual can be used by anyone — experienced mechanics, shop trainees, vocational students, and lay readers.

Persons not familiar with the topics discussed in this book should begin with Chapter 1 and then study the chapters in sequence. The experienced person can find what is needed on the "Contents" page.

Each guide was written by Deere & Company, John Deere Publishing staff in cooperation with the technical writers, illustrators, and editors at Almon, Inc. — a full-service technical publications company headquartered in Waukesha, Wisconsin (www.almoninc.com).

FOR MORE INFORMATION: This book is one of many books published on agricultural and related subjects. For more information or to request a FREE CATALOG, call 1–800–522–7448 or send your request to address above or:

**Visit Us on the Internet**
**http://www.johndeere.com/ publications**

ACKNOWLEDGEMENTS:

Contributing Authors and Editors:
Frank Buckingham
Arland W. Pauli
Harold Thorngren
Bruno Johannsen
Vernon D. Hagelin
Roland F. Espenschied
Thomas A. Hoerner, PhD.
Keith R. Carlson
Stewart W. Melvin, Ph.D.
Ralph Reynolds
John Deere gratefully acknowledges help from the following people and groups:
David Schmerse
Glenn Olson
Tom Bueker
Tim Claus
And a host of other John Deere people:
Fleischer Mfg. Inc., Columbus, NE.
Hiniker Company, Mankato, MN.
Richard R. Johnson
John C. Siemens
K Olson
University of Illinois
Soil Conservation Service Des Moines, IA.
DMI Inc., Goodfield, IL.
M&W Gear, Gibson City, IL.
Fuerst Brothers, Rhinebeck, NY.
Krause Plow Corp., Hutchinson, KS.
Sunflower Mfg. Co., Beloit KS.
Richardson Mfg. Co., Cawker City, KS.
Working Tires, Johnson Hill Press Inc. Editorial Office, Fort Atkinson, WI.
Almon, Inc. Waukesha, WI.

We have a
long-range interest in
Vocational Education


ISBN 0-86691-344-0


12/16/10

## 5 TOOLBARS



## 6 PRIMARY TILLAGE

## 7 SECONDARY TILLAGE

# 8 DRYLAND TILLAGE



# 9 TILLAGE AT SEEDING



# 10 POST-EMERGENCE TILLAGE



# 11 APPENDIX

# 12 ANSWERS TO TEST YOURSELF QUESTIONS



# 13 TILLAGE HISTORY



# GLOSSARY

# INDEX

# Tillage Systems and Conservation Planning

# 1

JDPX6061

*Fig. 1 — A Plow Used When America's Heartland Was Settled*

## Introduction

Tillage has been defined as those mechanical, soil–stirring actions carried on for the purpose of nurturing crops. The goal of proper tillage is to provide a suitable environment for seed germination, root growth, weed control, soil–erosion control, and moisture control—avoiding moisture excesses and reducing stress of moisture shortages.

Tillage requires well over half of the engine power expended on American farms, and it has been estimated that more than 250 billion tons of soil are tilled each year in this country. Many of the implements used, and much of the need for all this soil movement, have long been taken for granted. There actually have been few revolutionary developments in tillage equipment. Most changes have been the result of modification, improvement, and evolution of earlier equipment designs and ideas. However, there have been more changes in tillage implements and methods in the last 100 years than in previous recorded history, and more growth in mechanization in the last 25 years than in the preceding century (Fig. 1, Fig. 2, and Fig. 3).

Some changes have been quickly accepted, but others died in infancy. Nevertheless, the continued change and growth in equipment and practices have helped the American farmer become the most efficient food producer in history.

The primary objective of any cropping program is continued profitable production, so most farmers prefer to follow proven practices with readily available equipment. This offers reasonable assurance of predictable results with least risk.

But no tillage operation can be justified merely on the basis of tradition or habit. Any tillage practice that doesn't return more than its cost by increasing yield and improving soil conditions should be eliminated or changed. Contrary to previous beliefs, soil needs to be worked only enough to ansure optimum crop production and weed control. Any tillage activity beyond that is of questionable value.

JDPX6137

*Fig. 2 — Huge Steam Tractors Pulled Large Gang Plows During the Late 1800s*

JDPX6138

*Fig. 3 — Motorized Row–Crop Cultivator During the Early 1900s*

**Tillage Systems**

Basic concepts and objectives of tillage remain largely unchanged. However, major changes in tillage systems have occurred since the early 1980s. Tillage equipment has been developed or reconfigured to meet requirements of these new systems.

The greatest change in tillage systems has been a significant shift to conservation farming. This shift has occurred in response to concerns for reducing energy costs, soil erosion, fertilizer and pesticide use, water pollution, and operating costs generally. The trend to conservation farming has been intensified by federal legislation that requires conservation for crop support eligibility. USDA's Soil Conservation Service estimates that 25 percent of U.S. cropland is now under some form of residue management. The National Association of Conservation Districts

Conservation Technology Center estimated that conservation tillage was used on 40.7 percent of all 2004 planted acres in the United States. That continues a trend upward from 26.1 percent in 1990, 31.4 percent in 1992, 35.0 percent in 1994, 35.8 percent in 1996, 37.2 percent in 1998, 36.7 percent in 2000, and 36.6 percent in 2002 (Fig. 4). Legislation that requires conservation may increase conservation farming to half of the cropland.

Adoption by farmers of conservation systems has had a major impact on how much different types of tillage equipment are used. For example, Fig. 5 shows that industry shipments of moldboard plows and disks decreased to about 25 percent of the total (from 77 percent) between 1967 and 1990. At the same time, combination primary and secondary equipment, largely unavailable in 1967, constituted about 30 percent of shipments in 1990. Sales of chisel plows and field cultivators also increased to about 30 percent of all tillage equipment. These trends indicate greater adoption of tillage systems that reduce the amount of work done to the soil.

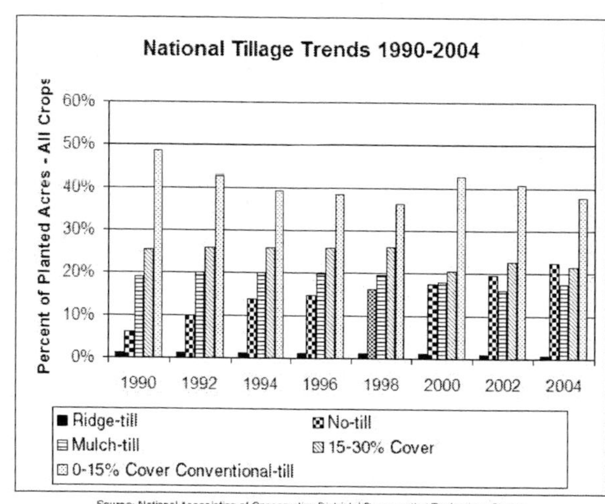


JDPX6068

*Fig. 4 — National Tillage Trends From 1990 to 2004*

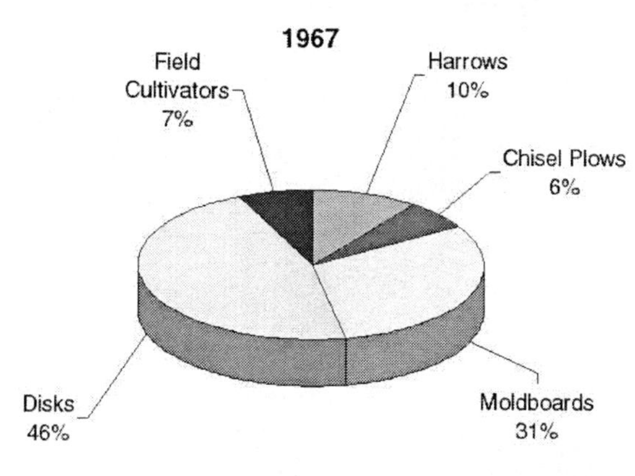


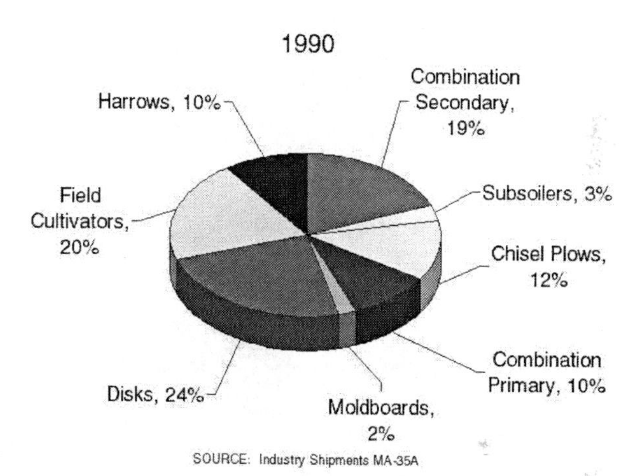


JDPX6066, 6067

*Fig. 5 — U.S. Tillage Industry Shipments, 1967 and 1990*

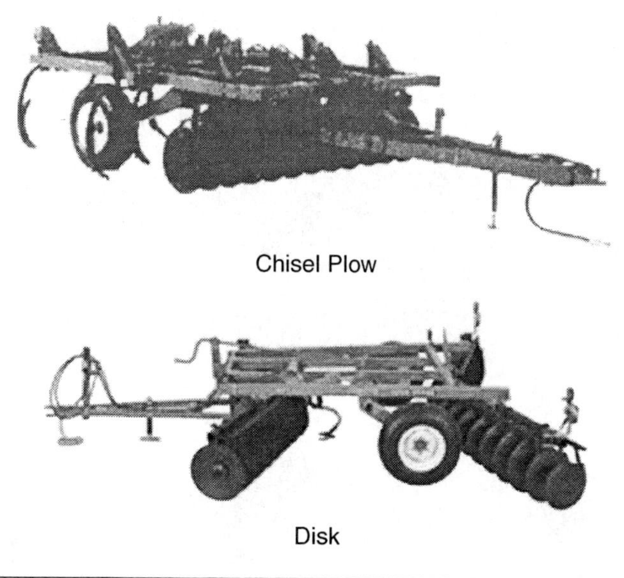

Chisel Plow

Disk

Subsoiler (or V–Ripper)

Moldboard Plow

JDPX6062, 6063, 6139, 6140

*Fig. 6 — Selected Primary Tillage Tools*

## Definition of Tillage Systems

Tillage can be defined as any mechanical manipulation of soil. Many different types of tillage tools are available to manipulate the soil. A tillage system is the sequence of tillage operations performed in producing a crop. For many tillage systems, the specific operations can be separated into:

- Primary tillage
- Secondary tillage

### Primary Tillage

Primary tillage is usually the deepest operation in the system. A deep tillage operation loosens and fractures the soil to reduce soil strength and to bring or mix residues and fertilizers into the tilled layer. The implements used for primary tillage include moldboard, chisel, and disk plows; heavy tandem, offset, and one–way disks; subsoilers; and heavy–duty, powered rotary tillers. A few different primary tillage tools are shown in Fig. 6. These tools usually operate at least 6 inches (15.4 cm) deep and produce a rougher soil surface than do secondary tillage tools. They differ from each other as to amount of soil manipulation and amount of residue left on or near the soil surface.

### Secondary Tillage

Secondary tillage is used to kill weeds, cut and cover crop residues, incorporate herbicides, and prepare a well–pulverized seedbed. Secondary tillage tools include light– and medium–weight disks, field cultivators, row cultivators, rotary hoes, drags, powered and unpowered harrows or rotary tillers, rollers, ridge– or bed–forming implements, and numerous variations or combinations of these. They usually operate at depths of less than 5 inches (12.7 cm). Some secondary tillage tools are shown in Fig. 7. Because of the number and variations of the tillage systems used by farmers, it is difficult to give each system a meaningful name or precise definition. The systems can, however, be identified or grouped according to one of the following:

- Overall objective: Names of systems include conservation tillage, conventional tillage, clean tillage, minimum tillage, reduced tillage, and mulch tillage.
- Specific operation: Names of systems include moldboard plow, chisel plow, disk, no–till or no–tillage, and ridge–till.

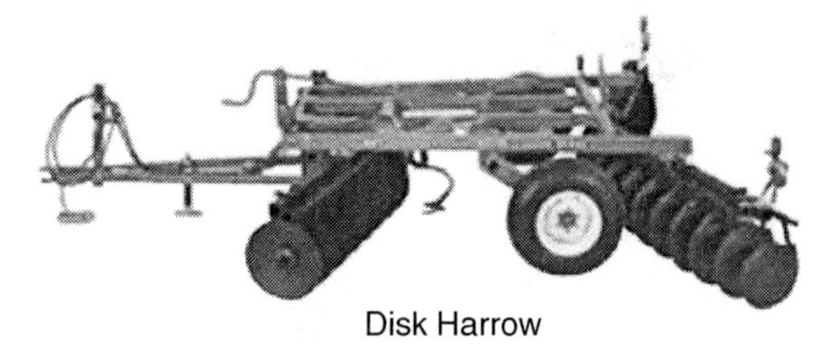

Disk Harrow

Field Cultivator

Rotary Hoe

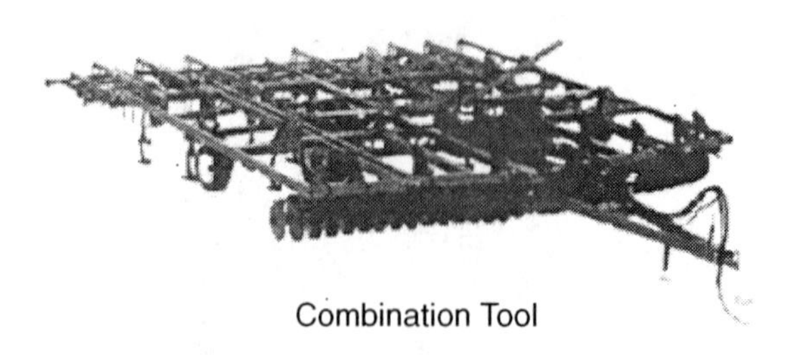

Combination Tool

Row–Crop Cultivator

JDPX6062, 6064, 6065, 6141, 6142

*Fig. 7 — Selected Secondary Tillage Tools*

The name problem is further compounded by the fact that definitions differ between regions of the country. Different names are used to mean the same tillage system, or the same name refers to different tillage systems, all depending on where you live. Therefore, to accurately define or describe a given tillage system, all the operations that make up the system should be listed. The list should include all the tillage operations, any chopping or shredding of residue, application of pesticides and fertilizers, planting, cultivating, and harvesting.

Recognizing the confusion in terminology, here are the general definitions of the more common tillage systems grouped according to overall objective and specific operation.

## Tillage Systems Named by Overall Objective

### Conservation Tillage

Conservation tillage is defined as the field operations required for profitable crop production while minimizing soil erosion due to wind and water. The emphasis is on soil conservation; but conservation of soil moisture, energy, labor, and even equipment are sometimes additional benefits.

To be considered conservation tillage, a system must produce, on or in the soil, conditions that resist the erosive effects of wind, rain, and flowing water. Such resistance is achieved either by protecting the soil surface with crop residue or growing plants, or by increasing the surface roughness or soil permeability.

Conservation tillage is often defined as any crop production system that provides one of the following:

- A residue cover of at least 30 percent after planting to reduce soil erosion due to water (Fig. 8).
- At least 1000 pounds per acre (1121 kg/ha) of flat, small grain residues, or the equivalent, on the soil surface during the critical erosion period to reduce soil erosion due to wind.

Conservation tillage generally represents a broad spectrum of tillage and planting systems. Depending on the preceding crop, examples of conservation tillage includes any one of the following specific systems:

- Chisel plow or subsoil
- Blade plow
- Disk and/or field cultivate
- Strip–till
- Ridge–plant
- No–till

The above specific systems are defined in the next section. There may be other conservation tillage systems you know about, and others will undoubtedly be developed.

### Conventional Tillage

Conventional tillage refers to the sequence of tillage operations traditionally or most commonly used in a given geographic area to prepare a seedbed and produce a given crop. The operations used vary considerably for different crops and from one region to another and even within a region.

JPDPX6172

*Fig. 8 — A Subsoiler Being Used in a Conservation Tillage System*

*Fig. 9 — A Field Being Planted With a Conventional or Clean Tillage System*

For instance, on the prairie soils of central Illinois, conventional tillage for corn means this: in the fall, phosphorus, potassium, and lime may be applied and the soil is chisel plowed; in the spring, the ground is disked, nitrogen fertilizer and herbicides are applied before field cultivating once or twice, and the crop is planted. The growing crop is then rotary hoed if necessary and cultivated once. On the other hand, for non–irrigated corn production in Nebraska, conventional tillage consists of the following sequence of operations, all done in the spring: fertilizer application, disking, herbicide application during a second disking, and field cultivating before planting, with the crop subsequently cultivated one or two times. Usually, with conventional tillage, only a small amount of residue (less than 15%) is on the soil surface at planting (Fig. 9).

### Clean Tillage

Clean tillage involves a sequence of operations that prepares a seedbed having essentially no plant residues on the soil surface (Fig. 9). Many conventional tillage systems are also clean tillage systems, particularly those that include use of the moldboard plow. A soil surface essentially free of residue can also be achieved with other implements, such as the disk, especially after crops that provide fragile residues like cotton or soybeans.

### Minimum Tillage

The term *minimum tillage* has been used for many years. The definition has varied over the years and from region to region. The term is not very meaningful. Perhaps this is the best definition: the minimum soil manipulation necessary for crop production or for meeting tillage requirements under existing conditions. When most people use the term *minimum tillage*, they mean reduced tillage as defined below.

### Reduced Tillage

Reduced tillage refers to any system that is less intensive and less aggressive than conventional tillage. Either the number of operations is decreased, or a tillage implement that requires less energy per unit area is used to replace an implement typically used in the conventional tillage system. Because it is not specific, the term *reduced tillage* is also not very meaningful.

### Mulch–Till

Mulch–till (Fig. 10) is any conservation tillage system which includes the use of implements that till the entire soil surface. Prior to planting, tillage implements such as a chisel plow, blade plow, rod weeder, disk, or field cultivator are used. At least 30 percent of the soil surface must be covered with residue after planting. Weed control is accomplished with herbicides and/or row cultivation.

*Fig. 10 — A Disk Used in a Mulch–Till System*

## Tillage Systems Named by Specific Operation

Several tillage systems are named by the major tillage implement used. Some of these are:

- Moldboard plow
- Chisel plow
- Subsoiler
- Blade plow
- Disk
- Field cultivator
- Combination tool

### Moldboard Plow, Chisel Plow, and Subsoiler Systems

The moldboard plow, chisel plow, or subsoiler (Fig. 11) is used for primary tillage, most often in the fall. Various secondary tillage operations are performed prior to planting.

Before primary tillage, the previous crop residues are sometimes chopped or disked, and fertilizers and lime may be applied. Primary tillage implements like the chisel plow or subsoiler generally leave a rougher soil surface and a considerably greater percentage of the residue on or near the soil surface than does the moldboard plow.

Secondary tillage varies widely in the type and number of operations. If herbicides are to be incorporated with tillage, the fall–tilled soil is often leveled in the spring with a disk or field cultivator. Herbicides are then applied and incorporated with one or two passes of a tillage implement like a disk, (Fig. 12), field cultivator, or combination tool (Fig. 13). Fertilizers, especially anhydrous ammonia for corn, can be applied in the fall, in the spring after the leveling operation, or after planting. A rotary hoe and a row–crop cultivator are frequently used for additional weed control and soil loosening after crop emergence.

If herbicides are not incorporated with tillage, one or more secondary tillage operations are used to level the soil and prepare a seedbed. Herbicides are applied after planting. Fertilizers may be applied before, during, or after planting.

**Blade Plow System**

In the Great Plains, the blade plow is used for primary tillage after harvest of small grains. A blade plow has V–shaped blades which are 3 to 5 feet (0.91 to 1.52 m) wide and operated at a relatively shallow depth. A blade plow typically leaves more residue on the soil surface than a chisel plow. Secondary tillage may consist of three to five rod weeding or harrowing operations for weed control before planting the subsequent crop. Fertilizer and herbicide applications are similar to the chisel plow and moldboard plow systems. This system is called stubble–mulching in some regions.

JDPX6060

*Fig. 11 — A Chisel Plow Being Used for Primary Tillage*

JDPX6226

*Fig. 12 — A Disk Being Used to Prepare a Seedbed*

**Disk, Field Cultivator, or Combination Tool Systems**

Tillage is not as deep with these systems as with moldboard plow, chisel plow, or subsoil systems. The number and types of operations in the disk and field cultivator systems vary considerably. Primary tillage, if used, consists of using an offset or tandem disk (Fig. 12), field cultivator, or combination tool (Fig. 13) to incorporate herbicides and prepare a seedbed. In some conditions, only one of these secondary tillage tools is used. Thus, a seedbed is prepared and herbicides are incorporated in a single operation.

JDPX6117

*Fig. 13 — A Combination Tool Is Commonly Used to Incorporate Herbicides and Prepare a Seedbed*

One innovative implement for use in conservation tillage systems (Fig. 14) is capable of tilling through heavy residue and leaving a high percentage of the residue on the soil surface to control erosion. It is also claimed to provide thorough herbicide incorporation and a weed–free seedbed in one secondary tillage pass.

JDPX6173

*Fig. 14 — An Implement for Conservation Tillage*

### No–Till (Also Sometimes Called Ero–Till or Slot–Plant)

The soil is left undisturbed from harvest to seeding and from seeding to harvest (Fig. 15). The only tillage is the soil disturbance done by the planter or drill. Coulters mounted on the planter or drill till a narrow strip or slot into which the seed is planted. On some planters and drills, coulters are not used; the seed opener is used to cut the residue and till the narrow strip.

Strictly speaking, the planting or drilling operation is the only operation used that disturbs the soil. In some cases the basic no–till system is modified by using of a drag harrow, rotary hoe, row cultivator, or knife fertilizer application.

When such tools are used in an otherwise no–till system, the modification should be noted.

JDPX6145

*Fig. 15 — A Crop Being Planted With the No–Till System*

### Ridge–Plant (Also Sometimes Called Ridge–Till or Till–Plant)

The soil is generally left undisturbed from harvest to planting except possibly for fertilizer injection. Crops are planted and grown on preformed ridges (Fig. 16). The ridges are usually formed in the previous crop either when cultivating or for furrow irrigation when hilling or ditching. Typically, ridges are re–formed annually when cultivating, hilling, or ditching (Fig. 17). The height of the ridges should be 6 to 10 inches (15 to 25 cm). For erosion control, the ridges should be at least 3 inches (7.6 cm) higher than the furrows after planting.

JDPX6146

*Fig. 16 — Planting With the Ridge–Till System*

In most ridge–till systems, the planter is equipped with sweeps, disk row cleaners, coulters, or horizontal disks (Fig. 17). These row cleaning attachments remove about 1 to 3 inches (2.5 to 7.6 cm) of soil, surface residue, and weed seeds from the row area. Ideally, this leaves a residue–free strip of moist soil on top of a ridge into which the crop is planted.

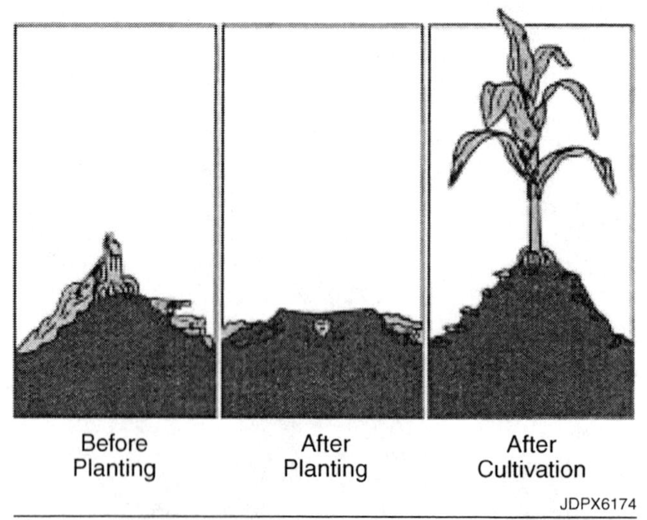


JDPX6174

*Fig. 17 — With the Ridge–Till System, the Ridges Are Rebuilt During Cultivation*

## Eliminating Tillage Operations and Improving Field Efficiency

Major objectives of conservation tillage include eliminating operations and increasing surface residue at planting. Maximum field efficiency is an important consideration in selecting equipment to meet these objectives.

Reducing the number of field operations saves fuel and labor costs, usually permits earlier planting, and can lower equipment investment if some tools are eliminated. Reduced wheel traffic can also reduce soil compaction. The amount of fuel and labor saved depends on the number of trips, before and after, over the field and the type of tillage operations eliminated. For example, eliminating plowing saves more fuel than cutting a harrowing operation. Switching from conventional to zero tillage can save about 75 percent of the tillage energy; however, part of that savings may be offset by the cost of energy to produce extra chemicals for weed and insect control.

Field capacity can be, either theoretical field capacity or effective field capacity. Theoretical field capacity (TFC) is the area that could be tilled in one hour if no time were lost to actions such as turning at the ends of the field or adjustments or repairs.

TFC (Acres per hour)

$$TFC = \frac{\text{Width (ft)} \times \text{Speed (mph)}}{8.25}$$

TFC (Hectares per hour)

$$TFC = \frac{\text{Width (m)} \times \text{Speed (km/h)}}{10}$$

Effective field capacity (EFC) is the amount of work various tools can accomplish while operating in actual field operations. This value is determined by time and motion studies or simply by observations in the field.

Field efficiency (FE) is the ratio of the area an implement can theoretically cover in one hour (TFC) to the area actually covered (EFC). The relationship is expressed as a percentage of TFC by the formula:

$$FE = \frac{EFC}{TFC} \times 100$$

Field efficiencies of most tillage tools range from 70 to 90 percent. Factors that influence field efficiency include organization and size of equipment, travel speed, shape of the field, condition of equipment, condition of the soil, and the experience and health of the operator. For a detailed discussion on selection and use of equipment for maximum field efficiency, see FMO Machinery Management.

## Effects of Conservation Farming Systems

A conservation system can have highly variable effects. A mulch of crop residue reduces moisture evaporation and surface runoff. Erosion–prone soils are thus prime candidates for a conservation system. Tillage that maintains crop residue but leaves the surface rough may reduce water runoff over that of a zero–tillage system.

Compared with conventional tillage, conservation tillage can require more careful crop management of soil fertility and control of weeds, insects and diseases. For instance, lime, phosphorus and potassium are not highly mobile in the soil because they tend to remain near the surface if not incorporated by tillage or banded below the surface, increased application rates may be required. Also, rough, trashy surfaces produce an unfavorable environment for germination of weed seeds, but make effective herbicide application and mechanical cultivation more difficult. Surface–germinating and perennial weeds become more difficult to control as tillage is reduced. But most insect and disease problems that might be increased by conservation farming can be overcome by using resistant crop varieties or applying pesticides.

Reducing the amount of tillage can save time at planting and also reduce fuel, labor and equipment costs. But some of these savings can be offset by increased pesticide and fertilizer costs.

Crop yield responses to conservation systems differ with location and soil type. Because rough, trash–covered soil warms more slowly in the spring, reducing tillage may increase yields on warm drought–prone soils or on soils subject to compaction problems. Well drained soils that dry and warm rapidly often produce similar yields over broad ranges in tillage systems. But poorly drained soils, especially in cooler regions, often yield less as tillage is reduced. Wet conditions often delay timely field operations and reduce early season crop growth. University and soil conservation personnel can provide information on tillage systems for specific soil types.

## Developing a Plan for Conservation Farming

Although farmers favor the idea of conserving resources, many things must be considered before deciding on a specific method of conservation farming. A change in one practice may have a significant effect on many other operations. Also, what is best for one farm may not be economical for another.

Good management involves selecting the best system for particular soil and climatic conditions. Selection and operation of equipment is also important, as is having an understanding of, and faith in, the system. For example, if seeing trash on the surface before planting and cultivating is worrisome, the farmer may reject the tillage system. Some

farmers say that the main obstacle is changing traditional attitudes about what has to be done and how it should look

Selection of a conservation farming system should be considered in the context of the total farm system. A conservation practice may greatly affect the operation of part of the farm, but the benefit could be small when compared to the entire farm unit.

Specific steps that may be followed to select a conservation system and put it into operation include:

- Evaluate the existing situation.
- Establish overall performance goals.
- Identify specific improvements.
- Identify available alternatives.
- Select conservation practices.
- Evaluate progress.
- Restate or modify goals.
- Repeat steps as needed.

### Evaluate the Existing Situation

Develop a picture of the farm and the movement of materials onto and away from the farm. Collect other information such as soil types, slopes, data on weather, crop production schedules, machinery inventory, current conservation activities, fertilizer application rates, pest problems, soil fertility and pH, and similar information. Major problems such as areas subject to water or wind erosion should be noted. Use this information to establish the rest of the plan. Compare results against this base information to evaluate progress.

### Establish Overall Performance Goals

Examples of performance goals:

- Soil erosion not to exceed 4.2 tons (9.4 t/ha).
- Energy consumption to decrease 5% in first year.
- Fuel consumption to decrease 7% in one year.
- Crop yields to be targeted at equal or higher levels.
- Other goals for pesticide use, fertilizer efficiency, etc.

### Identify Specific Improvements

Items that need specific action are individual pieces of equipment, production processes or items that are clearly inefficient. Specific improvements are easy to evaluate. They are also an easy way to conserve, since they obviously guarantee improvement if they are given attention.

### Identify Available Alternatives

Several alternatives may be available to achieve performance goals. For example, energy conserving alternatives may include reducing fuel consumption by cutting the number of trips over each field, reducing the amount of crop drying after harvest, or switching to alternate energy sources, if possible, for some operations.

### Select Conservation Practices

Evaluate individual conservation practices. Select a combination of practices and complementary equipment that gives promise of achieving goals.

### Evaluate Progress

Evaluate the total program annually. Review the list of improvements to see if they have been completed. Check gross performance goals.

### Restate or Modify Goals

If one of the improvements has been completed, remove it from the list. If a goal has been reached, restate it and make further improvements.

### Repeat Steps as Needed

Each year compare progress and change the system, if necessary, to make further improvements. The result will be a steady series of improvements.

## Conservation Planning

The charts that follow are designed to stimulate the kind of thought processes that should precede any decision to adopt a specific conservation system. Although chart entries are specific, they are not recommendations. They are designed only to depict some of the ways conservation practices may interact with other practices and farm operations. Consult experts in conservation planning for specific recommendations on a farm.

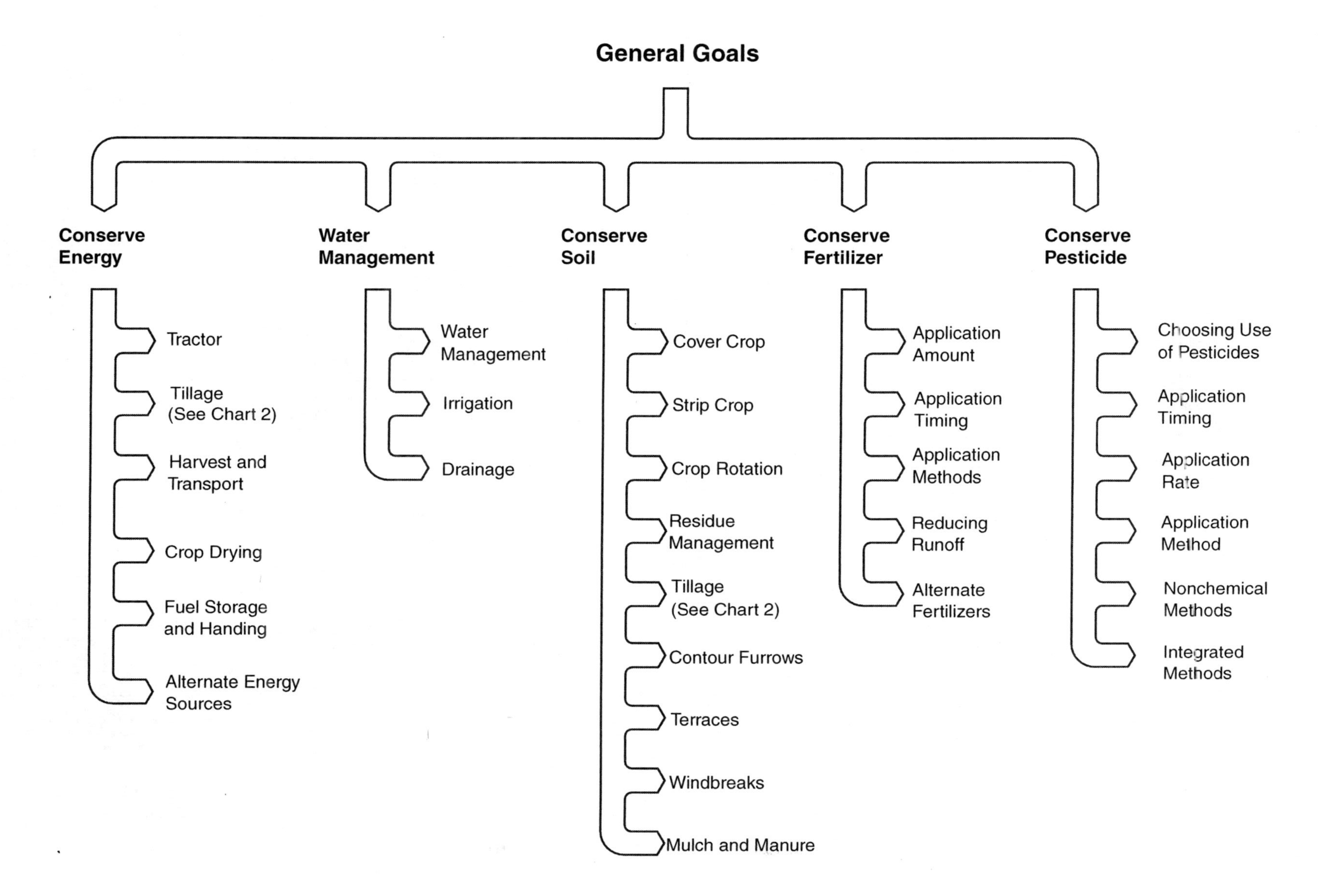
General Goals
Conserve Energy
Tractor
Tillage (See Chart 2)
Harvest and Transport
Crop Drying
Fuel Storage and Handing
Alternate Energy Sources
Water Management
Water Management
Irrigation
Drainage
Conserve Soil
Cover Crop
Strip Crop
Crop Rotation
Residue Management
Tillage (See Chart 2)
Contour Furrows
Terraces
Windbreaks
Mulch and Manure
Conserve Fertilizer
Application Amount
Application Timing
Application Methods
Reducing Runoff
Alternate Fertilizers
Conserve Pesticide
Choosing Use of Pesticides
Application Timing
Application Rate
Application Method
Nonchemical Methods
Integrated Methods
JDPX5909

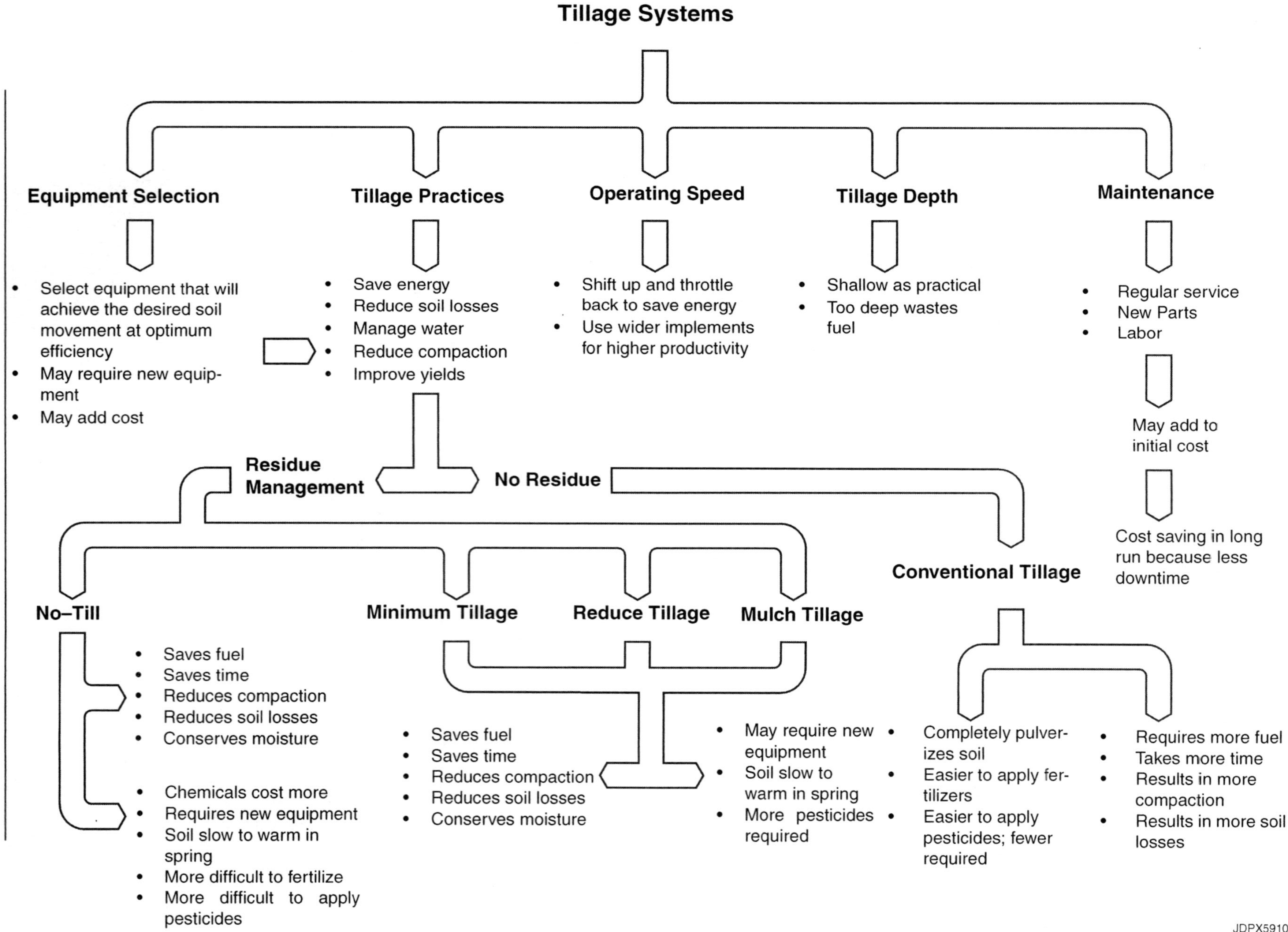
Tillage Systems
Equipment Selection
Tillage Practices
Operating Speed
Tillage Depth
Maintenance
• Select equipment that will achieve the desired soil movement at optimum efficiency
• May require new equipment
• May add cost
• Save energy
• Reduce soil losses
• Manage water
• Reduce compaction
• Improve yields
• Shift up and throttle back to save energy
• Use wider implements for higher productivity
• Shallow as practical
• Too deep wastes fuel
• Regular service
• New Parts
• Labor
May add to initial cost
Cost saving in long run because less downtime
Residue Management
No Residue
Conventional Tillage
No–Till
Minimum Tillage
Reduce Tillage
Mulch Tillage
• Saves fuel
• Saves time
• Reduces compaction
• Reduces soil losses
• Conserves moisture
• Chemicals cost more
• Requires new equipment
• Soil slow to warm in spring
• More difficult to fertilize
• More difficult to apply pesticides
• Saves fuel
• Saves time
• Reduces compaction
• Reduces soil losses
• Conserves moisture
• May require new equipment
• Soil slow to warm in spring
• More pesticides required
• Completely pulverizes soil
• Easier to apply fertilizers
• Easier to apply pesticides; fewer required
• Requires more fuel
• Takes more time
• Results in more compaction
• Results in more soil losses
JDPX5910

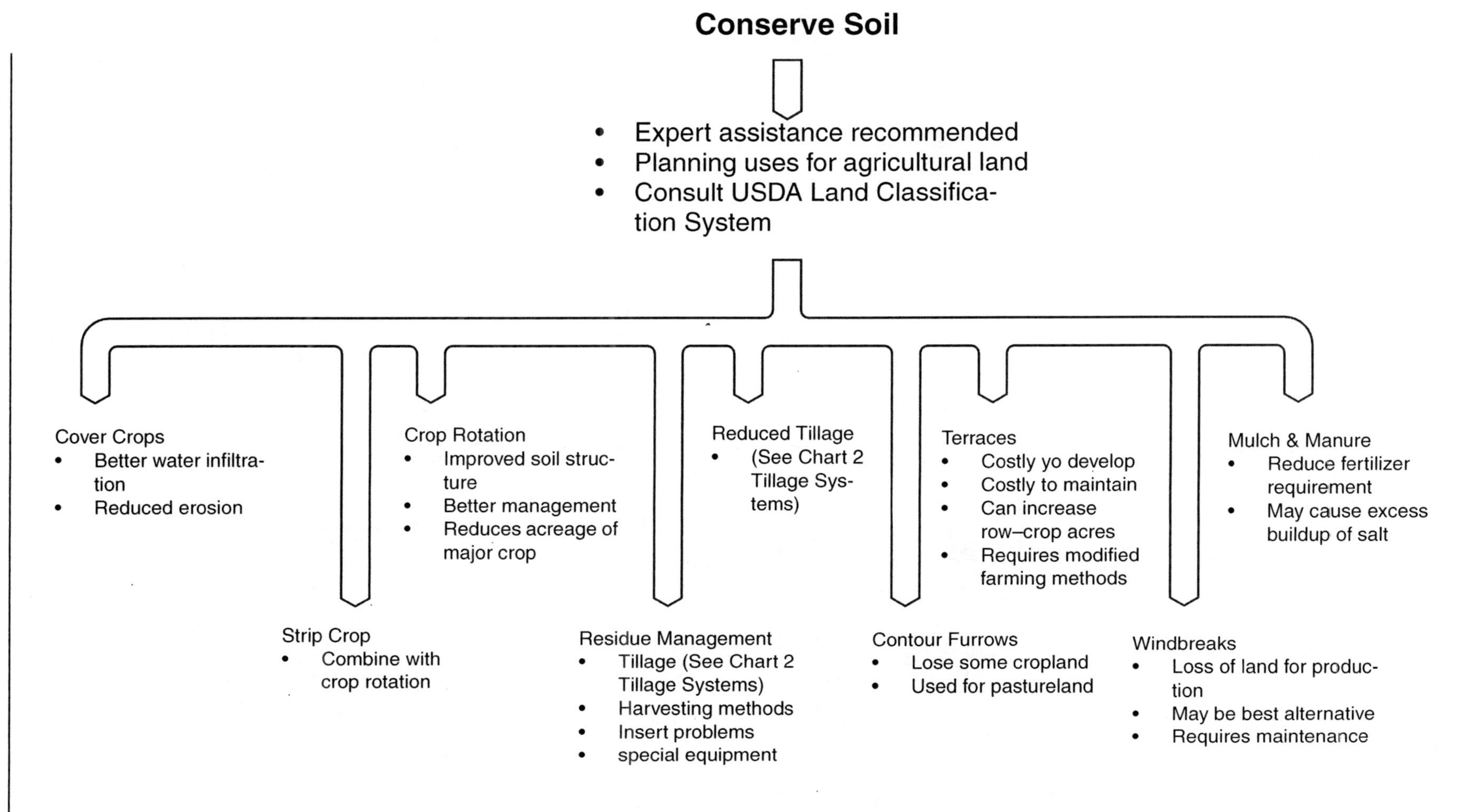
Conserve Soil
• Expert assistance recommended
• Planning uses for agricultural land
• Consult USDA Land Classification System
Cover Crops
• Better water infiltration
• Reduced erosion
Strip Crop
• Combine with crop rotation
Crop Rotation
• Improved soil structure
• Better management
• Reduces acreage of major crop
Residue Management
• Tillage (See Chart 2 Tillage Systems)
• Harvesting methods
• Insert problems
• special equipment
Reduced Tillage
• (See Chart 2 Tillage Systems)
Contour Furrows
• Lose some cropland
• Used for pastureland
Terraces
• Costly yo develop
• Costly to maintain
• Can increase row–crop acres
• Requires modified farming methods
Windbreaks
• Loss of land for production
• May be best alternative
• Requires maintenance
Mulch & Manure
• Reduce fertilizer requirement
• May cause excess buildup of salt
JDPX5911

## Summary

Regardless of the system used, more soil in the United States is over–tilled each year than is under–tilled. Over–tilling may result in excessive breakdown of soil–particle size, avoidable erosion, unnecessary compaction from wheel traffic, and wasted time and fuel.

Overworked seedbeds are a relatively modern phenomenon. When soil was tilled and crops were planted and harvested by hand, there was no time or energy to overwork the soil. Even with horses, there was little inclination to perform unnecessary tillage. With the introduction of tractor power, tillage became easier than ever before, and many farmers decided that if a little tillage was good, more would be better. They generally failed to recognize that overworking the soil could damage its structure and perhaps cause compaction.

A significant shift to conservation tillage has occurred since 1980 in response to higher energy, labor, and general operating costs. Soil erosion and water pollution concerns have also been a factor. The trend has been accelerated by legislation that mandates conservation for crop support eligibility. Adoption of conservation tillage systems has been accompanied by reduced numbers of tillage operations and a shift to tillage equipment that leaves soil rough and covered with surface residues from previous crops.

Selection of a conservation farming system should be considered in the context of the total farm system. Steps in developing and implementing a conservation plan are:

- Evaluate the existing situation.
- Establish overall performance goals.
- Identify specific improvements.
- Select conservation practices.
- Revise plan and repeat as necessary.

## Test Yourself

### Questions

1. Define tillage.
2. (Select one.) Tillage absorbs (a) one–fourth, (b) more than one–half, or (c) most of the power expended annually on American farms.
3. List six contributions of tillage to crop production.
4. (Fill in blanks.) ________ tillage cuts and shatters soil __________ tillage pulverizes and levels the surface prior to planting.
5. (T/F) Conventional tillage involves the same implements and operations for all crops and soil conditions.
6. What damage can result from overworking soil?
7. List at least three classifications of conservation tillage systems.
8. (T/F) With proper management, reduced tillage systems always provide higher crop yields.
9. List at least five steps in developing a plan for conservation farming.
10. (T/F) Conservation tillage systems, while numerous, always involve well–defined practices.
11. What is the primary goal of tillage?
12. What has been the greatest change in tillage systems since 1980?
13. (T/F) Sales of moldboard plows increased 25 percent from 1967 to 1990.
14. Define minimum tillage.
15. List three potential benefits and three possible detrimental effects from reducing the amount of tillage.

# Soil Compaction and Its Management

# 2

JDPX5851

*Fig. 1 — Root Growth in Compacted and Uncompacted Soil*

## Introduction

Much has been written and said about soil compaction (See Appendix, Suggested Readings). Soil compaction is often described as being entirely detrimental to crop yields. Actually, research has shown that, depending on circumstances, compaction can be either helpful or harmful. In this chapter, we will discuss what soil compaction is, what causes it, how crops respond to it, and some machinery considerations in developing farming systems to properly manage it.

## What Is Soil Compaction?

Soil scientists define compaction as a process of rearranging soil particles to decrease pore space and increase bulk density. An understanding of soil compaction requires that we first understand these terms and their relation to basic soil properties.

Soils contain particles of sand, silt, and clay. Sand particles can be of several sizes, but all sizes are much larger than silt particles which, in turn, are larger than clay particles. These soil particles, along with organic matter, make up the solid material of soil.

| | Size Range | |
|---|---|---|
| Soil particle | in. | mm. |
| Very coarse sand | 0.08–0.04 | 2.00–1.00 |
| Coarse sand | 0.04–0.02 | 1.00–0.50 |
| Medium sand | 0.02–0.01 | 0.50–0.25 |
| Fine sand | 0.01–0.004 | 0.25–0.10 |
| Very fine sand | 0.004–0.002 | 0.10–0.05 |
| Silt | 0.002–0.00008 | 0.05–0.002 |
| Clay | below 0.00008 | below 0.002 |

Pore spaces surround soil particles. During a heavy rain, water fills the spaces. After rain stops, water drains from the soil to a point where some water is held on the outside of soil particles and no further drainage occurs. This condition is called field capacity, i.e., the moisture–holding capacity of the soil after free drainage. At field capacity, air occupies the centers of the pores. A good rootbed has about 50% of the soil volume occupied by solid soil particles, 25% by water, and 25% by air. This is about the mixture in a loam soil with good granulation.

During compaction, soil particles are rearranged and packed closer together. The first pores to be lost are the large ones that help to increase water penetration and air storage. As soils become compacted, plant growth and

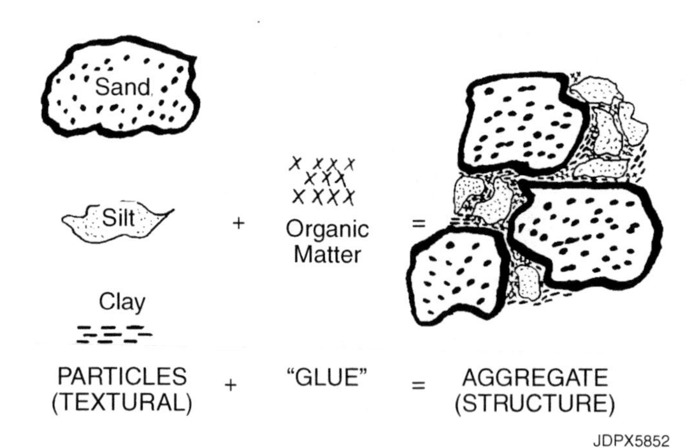


*Fig. 2 — Soil Structure*

productivity decline quickly when air space at field capacity is reduced and soil density is increased. For most important field crops, compaction problems begin if air space drops below 10 to 15% of total soil volume. Some soils become so dense that roots are restricted; they simply cannot physically penetrate the compacted zone, even if pore spaces contain air (Fig. 1).

## Importance of Soil Properties

Several soil properties affect the amount of compaction that will be developed in a given field or larger region. Soil structure (determined by organic matter and chemicals), soil texture, and soil moisture are all important. Not all of these factors can be easily changed. Yet, understanding how they influence soils can be very helpful in selecting or developing crop and machine systems to manage compaction.

### Soil Structure

Soil structure is the arrangement of soil solids into aggregates (Fig. 2). Organic matter aids formation of aggregates by binding sand, silt, and clay particles together. Organic matter acts as a glue that helps keep soil structure more stable. The result is a granulated structure with larger pore spaces between aggregates. Smaller pores exist within the aggregates. Because of organic matter's beneficial effect on soil structure, higher organic matter content can help make a soil more resistant to compaction. In contrast, high amounts of sodium reduce or prevent granulation. This causes soil particles to disperse and results in an undesirable structure. Such soils often become compacted, even where traffic does not occur. A soil layer high in sodium may develop a dense layer that restricts root growth and water drainage. If this layer is located too deep to till, the soils are difficult to drain and manage. Compounds of aluminum and calcium have also been associated with dense soil layers.

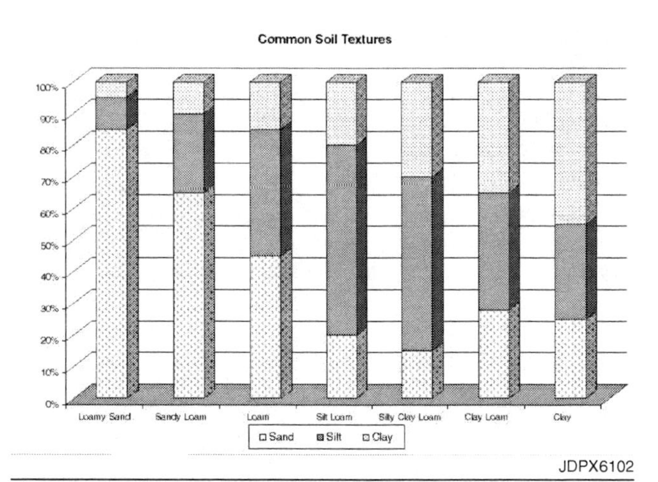


*Fig. 3 — Common Soil Textures*

## Soil Texture

The proportion of sand, silt, and clay in a soil greatly affects the ease of compacting a soil. Fig. 3 shows some common soil textures with typical proportions of sand, silt, and clay. Moderately coarse–textured sands, with enough organic matter to help provide structure, are the most resistant to compaction. Many loam, silt loam, and silty clay loam soils are also relatively resistant to compaction. As a general rule for loamy soils, increasing the clay content makes the soil more prone to compaction, particularly if organic matter contents are low. Soils classified as clays have more than 45% clay and are quite susceptible to compaction. They have a large proportion of pore space, but the pores are very small and remain filled with water. Soils with water–filled pores do not provide enough air for roots to grow well. Water–filled pores also may promote diseases and accelerate loss of nitrogen fertilizer. Soils with about equal proportions of different particle sizes are susceptible to compaction. The reason is that medium–sized particles fill in between coarse particles, and fine particles fill in between medium–sized particles. The result is very limited pore space. This is the same principle used to make concrete. Clay loams and sandy clay loams have about the same amounts of sand, silt, and clay; for this reason, they are more susceptible to compaction than other loamy soils. Likewise, even sandy soils with mixtures of sand sizes can be susceptible to compaction (Fig. 4).

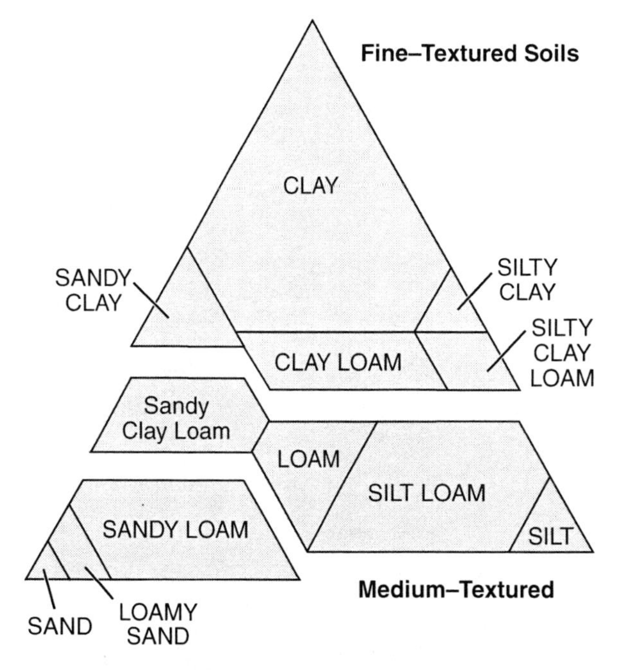


*Fig. 4 — Soil Textures and Available Pore Space (Compaction)*

Surprisingly, some low organic matter soils, such as loamy sands and sandy clay loams of the southeastern U.S., are among the most compaction prone soils. For reasons not fully understood, these soils often form a dense layer only a few inches thick somewhere in the upper 20 inches (51 cm). Some of these sands compact easily because a mixture of sand sizes fill up pore spaces between larger sand particles. Low organic matter, often less than 1%, and clay that does not shrink and swell as it dries and wets, also contribute to making these soils susceptible to compaction.

The dense layer in Southeastern soils is often called the plow layer or tillage pan. Tillage tools themselves are seldom the main cause of this dense layer. Rather, wheel traffic, animal traffic, or even rainfall provides the small force needed for such soils to compact. Layers above or below the pan can be of sand or some other texture not prone to compaction. This is why surface texture alone is not always a measure of a soil's compaction potential. Ripping a narrow slot through this dense zone will permit roots to reach water and nutrients in the subsoil. Ripping during bedding or planting is a common practice, but wheel traffic must be avoided above the ripped area or the dense zone will quickly re–form.

## Soil Moisture

Water in soil acts as a lubricant, making it easier for pressure to move soil particles close together. A moist soil compacts more easily than a dry soil. For example, only 5 psi (0.34 bar) of soil contact is required to compact a loam soil that's at field capacity (22% moisture) to 42% total pore space. With the same soil, at intermediate moisture, 48 psi (3.31 bar) of soil contact is needed, nearly 10 times as much pressure. And when the soil is dry, more than 160 psi (11.0 bar) is required. At all soil pressures, greatest compaction occurs at soil moisture near field capacity. Less compaction is caused when moisture is above field capacity because pore spaces begin to fill with water. A water–filled pore space can't be compacted!

During fall harvest and spring tillage seasons, most fields don't have crops actively growing and using water. As more and more rain falls, a field capacity wetting front moves downward from the soil surface. Large continuous pores from cracks, worm holes, or old root channels can cause variation in this wetting front.

In contrast, fields harvested under extremely wet conditions, resulting in wheel rutting, generally don't show effects of compaction because pore spaces are nearly water–filled.

Tilling, planting, and wheel traffic on soils at moisture levels above field capacity result in hard clods when dry, poor chemical incorporation, and poor soil tilth. Thus, there are good reasons, other than compaction, to avoid working soils above field capacity. Remember that, after a rain, a soil is most easily compacted if near field capacity or when the soil has dried out enough to work.

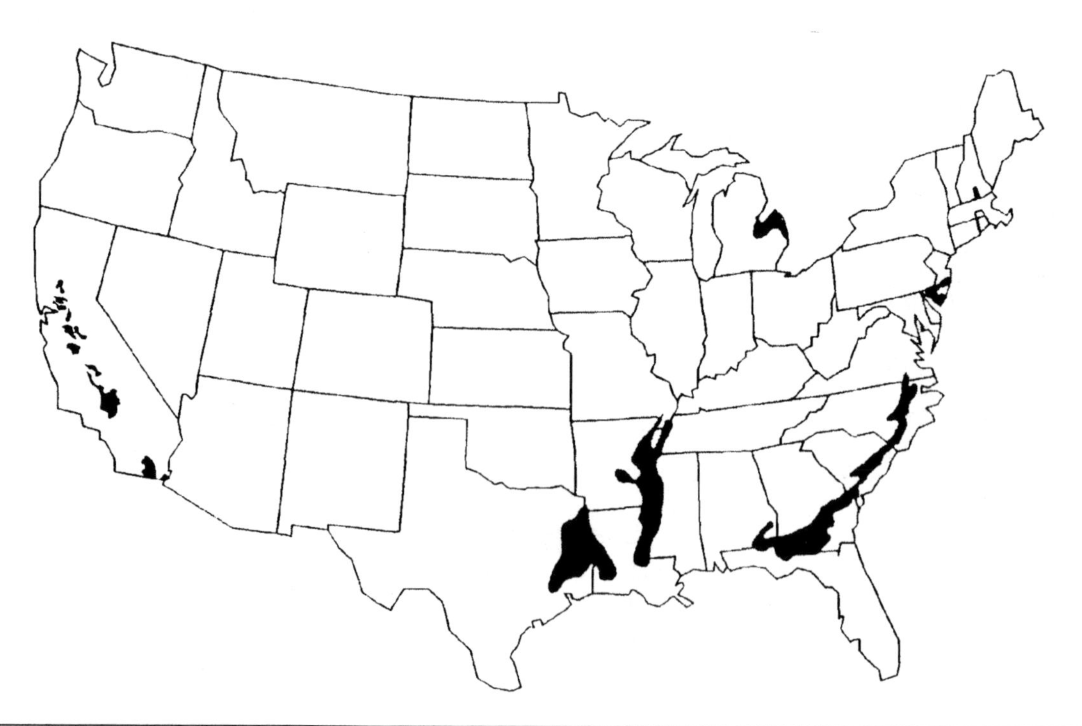


*Fig. 5 — Major Areas With Chronic Soil Compaction*

Soil moisture situations provide some tips on managing soil compactions:

- After harvest, conduct as many field operations as possible before soil is rewetted below the tillage depth.
- Consider fall tillage, particularly on non–erosive soils and those prone to compaction.
- During wet falls, confine traffic of heavy grain handling or fertilizer tender tanks to ends of fields or controlled traffic lanes.
- Schedule infrequent operations with heavy machines for dry seasons. Lime, fertilizer, or manure might best be applied to frozen ground if soil compaction is a problem.
- Conduct deep tillage, such as ripping, when the soil is dry enough to shatter. Wet soils don't loosen nearly as well.

### Major Chronic Compaction Regions

Areas where compaction is a natural soil condition are shaded in Fig. 5.

Not all have the same soil characteristics. For example, soils in the Mississippi Delta are high in clay; those in the Southeast have layers of loamy sands or sandy loams. Because of limited rainfall in the Great Plains, there are few problems in much of this region. The map does not show smaller problem areas that involve only a few hundred or few thousand acres. River bottom soils high in clay often are susceptible to compaction.

Many of the easily compacted soils shown in Fig. 5 are very productive and produce high value crops such as vegetables and cotton. Managing compaction must become a high priority in these regions.

## Crop Response to Compaction

Just as soil properties interact to affect a soil's susceptibility to compaction, several other factors interact to determine how a crop responds to compaction. These factors include:

- Amount of compaction
- Crop species and variety
- Growing season weather
- Fertilizer

### Management Optimum

#### Compaction

Soils can be compacted too much or too little for best results with a specific crop. Soil that is too loose may result in poor germination because of poor seed–soil contact, especially if dry. Evaporation loss may also be higher and upward movement of water from the subsoil is reduced if soils are too loose. In dry seedbeds, emergence is often better in compacted wheel tracks (Fig. 6). On the other hand, tracks in wet soils can easily become too dense, resulting in restricted root or seedling growth. This is why planters and drills usually have press wheels that can be adjusted to vary the amount of compaction. Dry seedbeds with little or no tillage require considerable firming pressure. As a general rule, firming pressure in moist seedbeds is adequate when the seed trench is closed and stepping on the area with the heel of your shoe causes only a slight added indentation in the soil surface.

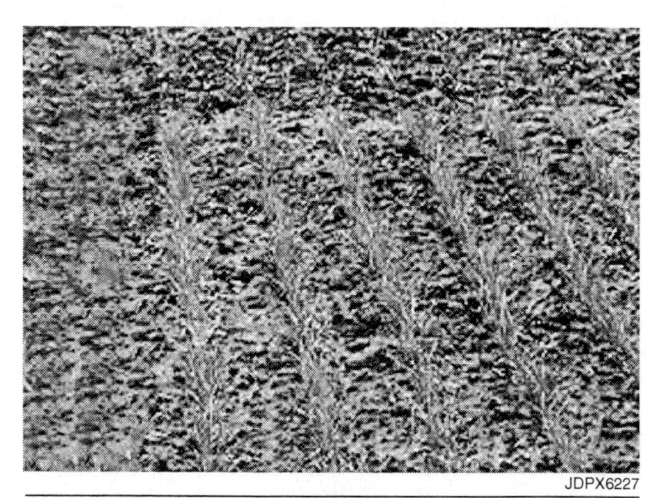

JDPX6227

*Fig. 6 — Improved Crop Emergence in Tractor Tire Tracks*

### Nitrogen Fertilizer Requirements

Crops requiring large amounts of nitrogen are sensitive to compaction because this fertilizer element is easily lost when soil becomes water saturated. In very wet soils, soil bacteria convert nitrogen to a gas that is lost to the atmosphere. Soil that is saturated with water can lose up to 70 pounds of nitrogen per acre (78 kg per ha).

Need for optimum compaction for higher yield remains throughout the growing season. A 13–year experiment in Minnesota, for example, showed that soybean yields were increased by wheel traffic compaction when May–to–August rainfall was less than 14 inches (36 cm). But compaction from wheel traffic reduced yields when rainfall exceeded that amount.

### Crop Species and Variety

Variety selection may be important with compaction–prone soils. Disease organisms, living in the soil, such as Phytophthora root rot are favored by moist conditions. Moderate compaction can reduce yields of soybean varieties that are not resistant to this disease. Resistant varieties are not affected by the same amount of compaction. Some dwarf small grain varieties can also be more sensitive to compaction because their germinating shoots do not become as long as those on taller varieties. This limits their ability to emerge from deeper, firmer soil conditions.

## Diagnosing Compaction Problems

The above has indicated what causes compaction and where its symptoms are most likely to occur. Remember: natural compaction is not a problem in most soil types, and effects in all soil types are not always negative. But, if a problem is suspected, keep the following in mind:

- Symptoms don't usually cover a whole field. They usually begin in wetter areas or where heaviest vehicles have traveled. An exception is the sandy soils of the Southeast, where symptoms may first show in droughty soils.
- Watch for standing water or areas that dry slowly. Check these later in the season to see if further compaction symptoms develop.
- Look for uneven plant heights or signs of plant stress.
- Watch for early yellowing that might indicate nitrogen loss in crops that require nitrogen fertilizer.
- Check to see if the soil has areas that contain few or no roots. Dig plants and check for odd–shaped or flattened roots (Fig. 7) or roots that grow horizontally or follow soil fracture planes.
- Don't jump to conclusions. Sometimes what appears to be symptoms of compaction can actually be caused by nutrient, diseases or herbicide stresses.

The best final indicator usually comes from digging plants midway in the season and checking roots. By this time, early disease and herbicide damage symptoms are usually past. Digging should be deeper than the tillage zone. If soil sections show root problems, poke with a knife or other pointed object to see if the soil is more dense than surrounding areas that have healthy roots. Visual inspection using a pointed object is usually better than using a penetrometer to measure soil resistance to pressure. Take note if problem areas are in a narrow layer or spread throughout the surface soil. Ripper and chisel shanks can break up compact layers, but moldboard plows are more effective in loosening entire surface horizons.

Finally, don't make management decisions based on a single piece of evidence. Some solutions can be costly. Consider all pieces of evidence and, if possible, follow suspect areas for more than one year. Sometimes the cause of a problem in one year or crop can turn into a benefit during another year or crop.

This variable response to compaction is illustrated by research at the University of Illinois (Table 1). A wet silty clay loam soil was compacted by 100% coverage with a 2–wheel drive tractor whose axle was weighted to 9 tons (8.2 t). The remaining area was worked with a tractor weighing less than 5 tons (4.5 t). During 4 years, the compaction had no effect on soybean yields. On average, corn yields were not affected; however, compacted areas yielded less during one two–year period and more during the other. In another similar test in Minnesota, when the wet clay loam soil was compacted with a 20–ton (18.0 t) load, it took 7 years for corn yields to return to normal levels.

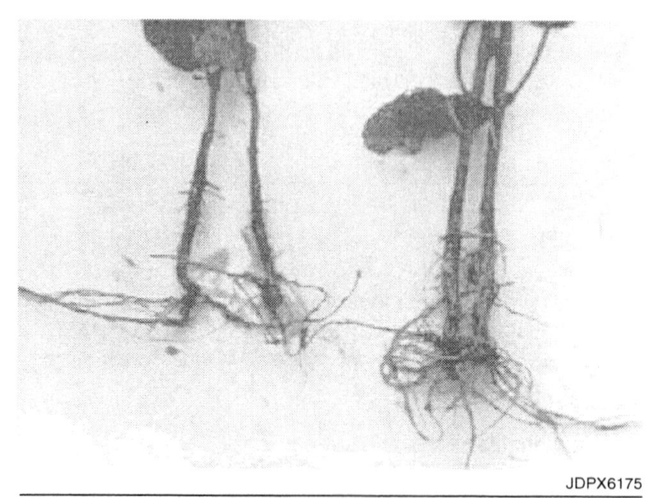

JDPX6175

*Fig. 7 — Flattened Root Caused by Compaction*

## Machinery Considerations

As farms have become larger and labor more costly, farmers have maintained productivity by using larger machines. As tractors become heavier, engineers have tried to keep tire soil contact pressures from increasing by using tires with larger diameters, wider widths, and tire construction that can operate at lower inflation pressures. Dual tires and tandem axles have also been added.

### Surface vs. Subsoil Compaction

Surface compaction is never deeper than the tillage layer of the top 8 to 12 inches (20 to 30 cm). Subsoil compaction occurs below the tillage layer. Surface compaction is caused by contact pressure applied at the soil surface; it depends mostly on tire inflation pressure. Below the tire, soil pressure (and compaction) decreases with depth. But compaction in deeper layers is determined by total load as well as contact pressure. Axle loads less than 5 tons (4.5 t) generally affect only the surface; loads above 10 tons (9.0 t) can also cause compaction below the normal tillage zone. It has been shown that loads of 20 tons (18.0 t) can compact a wet soil as deep as 2 feet (0.6 m). Subsoil compaction can persist for several years until released by natural factors. Freezing and thawing were once believed to be major factors in reducing subsoil compaction. But wetting and drying cycles as well as biological activity are now known to be equally or more important than freezing and thawing. Subsoil compaction can also be reduced by shrink–swell clays in some soils that cause cracks to develop as the soil dries. Deep–rooted crops such as alfalfa can also reduce subsoil compaction if the roots penetrate the dense soil zone.

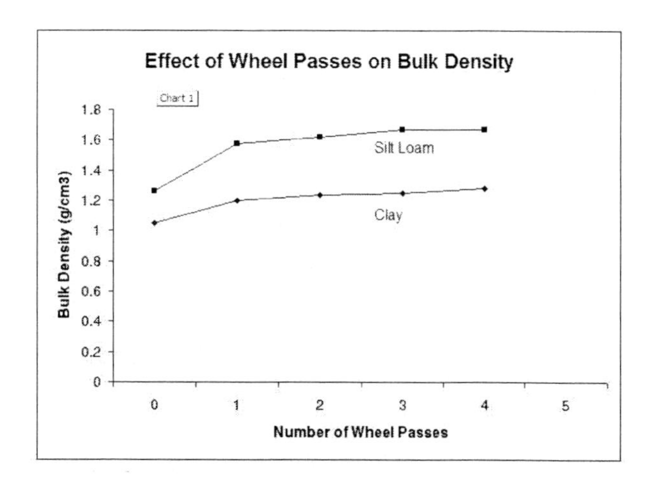


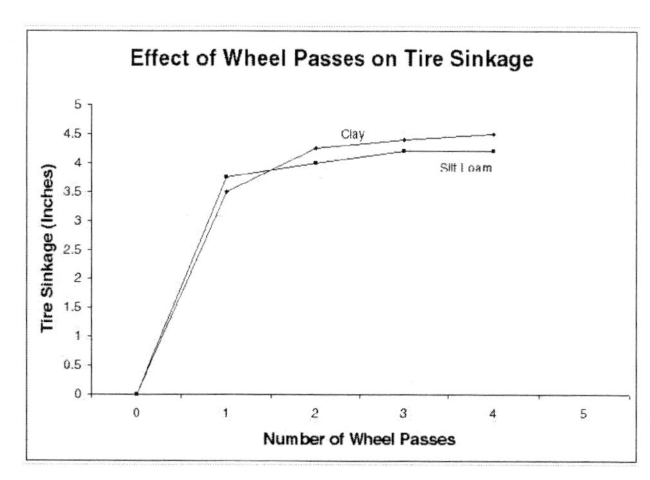


JDPX6103, 6104

*Fig. 8 — Effect of Wheel Passes on Soil Compaction*

| Year | No Extra Compaction Bu/A | No Extra Compaction (t/ha) | Compacted Bu/A | Compacted (t/ha) |
|---|---|---|---|---|
| 1986 | 170 | (10.7) | 184 | (11.5) |
| 1987 | 194 | (12.2) | 178 | (11.2) |
| 1988 | 93 | (5.8) | 100 | (6.3) |
| 1989 | 167 | (10.5) | 152 | (9.5) |
| Average | 156 | (9.8) | 154 | (9.7) |

*Table 1 — Variable Response of Corn to Soil Compaction*

### The First Wheel Pass Culprit

Tests have shown conclusively that the first pass of a wheel on loose, tilled soil will cause 80 to 90% of the total compaction caused by four passes (Fig. 8). If a wheel with a heavy load is followed by a wheel with a lighter load and similar tire pressure, no additional compaction occurs. On untilled soil, more passes may be needed to cause this degree of compaction, but the first few passes still have the greatest effect. This information can be used to manage compaction. Matching widths of front and rear wheels on a tractor helps. Matching tread widths of tractors and pulled transport vehicles such as grain carts and sprayers will also reduce field area subjected to tire surface pressures. Controlled traffic lanes can be used for successive field operations on soils susceptible to compaction. Wheel tracks are confined to the same path on each pass through the field. This system is easiest to use in ridge tillage, no–tillage, or bedded systems where track position is marked from one crop to another. It reduces area exposed to compaction and forms a lane that supports vehicles under less than optimum conditions. Its main disadvantage is the difficulty in optimum sizing of all machines for maximum productivity at least cost.

### Tractor Considerations

Modern tractors have chassis and ballasting options that can materially reduce compaction. A tractor chassis can be 2–wheel drive (2WD), mechanical front wheel drive (MFWD) or 4–wheel drive (4WD). Typical tractor ballast weight distributions are shown in Fig. 9. Compaction from larger tractors can be reduced by selecting a chassis that drives from the front and rear axles because all of the tractor weight is used for draft.

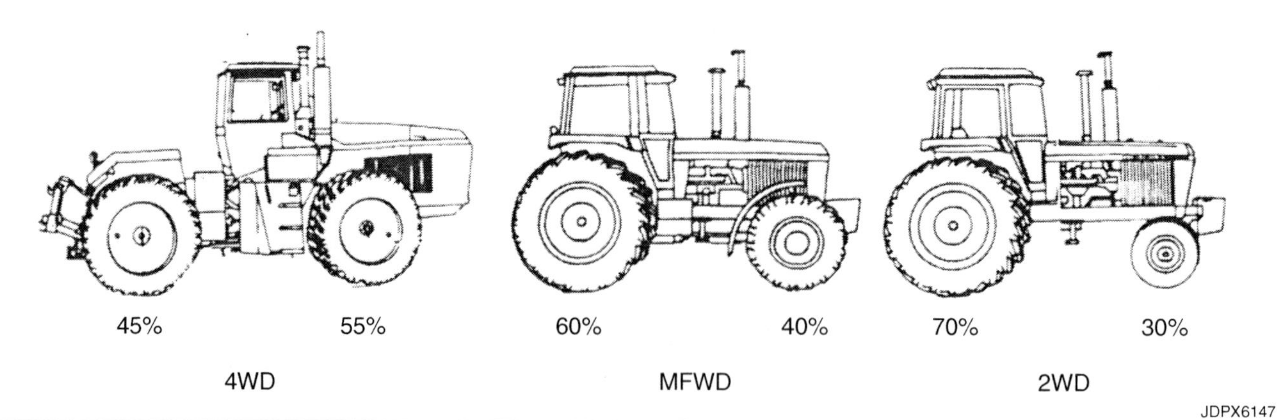


*Fig. 9 — Typical Tractor Ballast Weight Distributions*

| | PTO | | Weight per Axle | | | |
|---|---|---|---|---|---|---|
| | Hp | (kW) | Tons | (t) | Tons | (t) |
| 4WD | | | 45% Rear | | 55% Front | |
| | 325 | (242) | 10.2 | (9.3) | 12.5 | (11.3) |
| | 250 | (186) | 7.9 | (7.2) | 9.6 | (8.7) |
| | 200 | (149) | 6.3 | (5.7) | 7.7 | (7.0) |
| | 175 | (130) | 5.5 | (5.0) | 6.7 | (6.1) |
| MFWD | | | 60% Rear | | 40% Front | |
| | 200 | (149) | 8.4 | (7.6) | 5.6 | (5.1) |
| | 175 | (130) | 7.4 | (6.7) | 4.9 | (4.4) |
| | 150 | (119) | 6.5 | (5.9) | 4.2 | (3.8) |
| | 125 | (93) | 5.3 | (4.8) | 3.5 | (3.2) |
| 2WD | | | 70% Rear | | 30% Front | |
| | 200 | (149) | 9.8 | (8.9) | 4.2 | (3.8) |
| | 175 | (130) | 8.6 | (7.8) | 3.7 | (3.4) |
| | 150 | (119) | 7.4 | (6.7) | 3.2 | (2.9) |
| | 125 | (93) | 6.1 | (5.5) | 2.6 | (2.4) |

*Table 2 — Ballasted Tractor Axle Weights*

Table 2 shows tractor axle weights when ballasted at 140 pounds (63 kg) per PTO horsepower for heavy draft operation. Note that all but the largest 4WD tractors have maximum axle weights of about 10 tons (9 t) or less, the weight that can cause subsoil compaction. If high horsepower is needed on farms with compaction–susceptible soils, a common option is to use two MFWD tractors of 125 HP (93 kW) size. This will reduce maximum axle load to about 5 tons (4.5 t) per axle and should keep compaction shallow enough to be managed with routine tillage. Cost and labor trade–off are involved, but variations of this option are commonly used on compaction–prone soils of the southeastern U.S.

Tire options can also be used to increase contact area and reduce the potential for compaction. They include using dual instead of single tires, radial instead of bias tires, larger diameter or wider tires, and decreasing tire pressure. None of these options will reduce total load, but they will spread the load over a larger area, especially on firm soils.

Dual tires help reduce the depth that pressure moves into the soil. Duals on two–wheel drive row crop tractors also reduce wheel slip and increase ground speed under high draft loads. With MFWD tractors operated in front wheel assist, performance with single tires is about equal to duals; compaction will move slightly deeper, but single tires track only half as much of the field. With mounted implements, duals may be needed to carry the heavy load. Spaced duals are commonly used with combines in ridge tillage to allow narrow tires to go between rows.

**Tracked Vehicles**

Because they are efficient at spreading a load when bearing capacity of a soil is low, tracked vehicles have been thought to be a potential solution to soil compaction. Certainly they are effective when soils don't provide much support, such as in rice harvest or similar very wet conditions.

Recent research in Europe and the U.S. shows that both steel and rubber tracks actually are less efficient for limiting compaction in common farming situations where soils have reasonably good load bearing capacity but are sensitive to compaction. Several factors are involved, but the main reason is the uneven pressure distribution under tracks.

Pressures for tracked vehicles are commonly calculated by dividing total vehicle weight by total track area in contact with the soil. This assumes that pressure is equal over the entire track that contacts the soil. But, with a tracked vehicle calculated to create 5.7 psi (0.39 bar), pressure under the front idler wheel on the track started at 35 psi (2.41 bar) and increased as the cleat passed under each roller to a maximum of about 60 psi (4.14 bar) under the rear drive wheel. In the same field, the tire on a 4WD tractor never created more than 35 psi (2.41 bar) on the tread bar. Neither vehicle exerted much ground pressure in the area between bars on the tracks or wheels.

Other tests at several universities have also shown only minor differences in compaction under tracks compared with wheels. This is mainly due to uneven pressure distribution under tracks. On larger tractors, tracks have an advantage over dual wheels in contacting a narrower total width of soil. But this advantage must be weighed against the inability to vary width of track spacings, berming caused when turning at end of fields, and higher costs of tracked vehicles.

## Harvest and Transport Equipment

Tractors and tillage tools are often believed to be major causes of soil compaction. But the heaviest loads in fields are much more likely from harvest or transport machines. A loaded 500 bushel grain cart with single axle places a greater load on the soil than the largest tractor. Filled with corn or soybeans, it weighs more than 15 tons (13 t) or 50% more than the largest row crop tractor. Some of the largest grain carts have axle loads as high as 40 tons (36 t). Likewise, vehicles for fertilizer and pesticide application are often loaded at 10 tons (9 t) or more.

Many soils and crops will tolerate 10 ton (9 t) loads without yield reductions if combined with reasonable tillage, fertilizer, and cropping practices. But yield reductions are common when axle loads of 20 tons (18 t) or more are used on wet soils.

Since large machines are very important in farming, their use must be carefully planned. Combines can be equipped with wider tires or duals to limit compaction depth. They can also be unloaded at the ends of fields to limit the field area that is tracked. Grain carts with double axles will cause much less compaction, especially when they are operated in the path of combine wheels.

## Tillage Systems

The common belief that implements such as disks and plows cause compacted tillage pans is not supported by research. Excessive tilling can break down soil aggregates to make them more susceptible to compaction. But, used properly, tillage helps loosen soil and increases larger soil aggregates to help eliminate soil compaction problems. It is true that field traffic can cause compaction below the maximum tillage depth, but dense soil starts at the depth where loosening action of tillage stops. The tillage implement appears to cause the compact layer, but wheel traffic is, in fact, the cause.

Using tillage is one of the most important management options available to reduce surface compaction. In most northern and southern regions of the U.S., if soil compaction is present, moldboard plows are usually the most effective at loosening and improving soil tilth. Plowing inverts soil, bringing deeper soil to the surface where wetting and drying as well as freezing and thawing cycles can help restore good soil tilth. Less intense tillage not involving inversion is usually intermediate in reducing compaction.

Where plows are used to reduce compaction, heavy tractors should be operated on–land; a large tractor wheel in the furrow will cause a compaction zone beginning at plowing depth. On highly erosive soils, moldboard plowing is not advised. These soils should be deep–tilled with implements such as rippers or chisels equipped with sweeps to help preserve surface cover.

Soils that produce high yields with no–till systems generally have good internal and surface drainage and are not easily compacted. If no–till is used on soils prone to compaction, it's important to use crop rotation and to include occasional deep tillage in the system. If these soils are compacted below 12 inches (30 cm), rippers are one of the few options available. They are most effective when used in drier soils.

Combining field operations to reduce the number of trips over the field can help reduce compaction. Many tillage and planting machines have been developed to include machine components such as disk or coulter gangs, chisel sweeps, ripper standards, field cultivator sweeps, harrow attachments, and sprayers. These machines can perform operations in a single pass that required two or three passes with older equipment. Also, planters and drills can apply pesticides or fertilizers and sometimes handle minor tillage operations.

Some tillage and planting equipment can operate in and maintain surface residue. The mulch of surface cover left by this equipment is very effective in reducing soil erosion; but it often slows field drying. When combined with improved traction devices, conservation tillage equipment can operate in field conditions too wet for good tillage and planting. This makes it even more important to ensure that fields dry to working conditions before tillage to reduce compaction potential.

## Summary

Soil compaction results from pressure that rearranges soil particles to decrease pore space and increase soil density. It may be either natural or caused by farming practices. In some cases, it actually is helpful; in others, it is very detrimental to crop growth and yield.

Most soils are tolerant to widely used farming systems. Field traffic, not tillage, is the major cause of compaction, especially from operations that involve heavy loads when soil is wet. On most soils, axle loads less than 10 tons (9 t) will not cause subsoil compaction. Harvest and transport equipment have the heaviest loads. They can cause subsoil compaction that may persist for several years. Shallow compaction can result from lighter loads, but can be regulated with tillage and controlled traffic. A major objective in dealing with compaction should be to restrict it to the upper 8 to 12 inches (20 to 25 cm), where it can be managed with tillage.

Several options are available to manage compaction. It is very important to realize that a systems approach is required. Attempting to use a single machine, crop, or tillage system will often give disappointing results. With a crop and machinery management package, even soils very susceptible to compaction can be managed for profitable production.

## Test Yourself

### Questions

1. (T/F) Soil compaction is caused primarily by surface pressure from wheel traffic.
2. Define soil compaction.
3. What is soil structure? Soil texture?
4. List three categories of soil particles.
5. (T/F) Soil compaction is always detrimental to crop growth.
6. List at least three factors that affect crop response to soil compaction.
7. (Fill in blank.) The first pass of a wheel on loose soil causes __________ percent of total compaction from four passes.
8. What is the typical weight per tractor axle that can cause subsoil compaction?
9. (T/F) Tracked vehicles typically cause less soil compaction than wheeled vehicles.

# Crop Residue Management

3

## Introduction

Plant residues on the soil surface are very effective in reducing soil erosion. Erosion of topsoil begins when water detaches individual soil particles from the soil surface. The impact of a single raindrop may seem insignificant, but collectively, the raindrops during an intense storm strike the ground with tremendous force.

On bare sloping soil, two problems occur during intense rainstorms:

- Rainfall rate exceeds the infiltration rate of the soil.
- The infiltration rate decreases due to surface sealing.

When the rainfall rate exceeds the infiltration rate, water runs off the soil surface. As the infiltration into the soil decreases due to surface sealing, more runoff occurs. Runoff water can carry large amounts of soil with it. Thus, the potential for erosion of a bare sloping soil is great. A primary reason for using conservation tillage systems that leave crop residues on the soil surface is to reduce soil erosion. How effective is surface residue in reducing soil erosion?

## Surface Residues and Roughness Reduce Erosion

The effects of surface residue on soil erosion have been measured using rainfall simulators. At the University of Illinois a rainfall simulator was used to compare the water runoff and soil erosion of different tillage systems. The tillage systems included: 1) fall moldboard plow, 2) fall chisel plow, and 3) no–tillage. Rainfall simulator tests were made after growing corn and also after growing soybeans. The tests were made in the spring before any tillage or planting.

Fig. 1, top, shows rain being applied to a plot that had been used to grow corn the year before and moldboard plowed in the fall. The soil surface is exposed to the impact energy of the raindrops and the runoff and soil erosion are high.

The soil in Fig. 1, middle, had been chisel plowed in the fall after harvesting corn. A chisel plow with twisted shanks was used and the operation was performed on the contour. The ridges formed by the twisted points on the chisel plow are very effective in holding water and reducing the soil erosion. However, with enough rain the water will eventually break over the ridges or after the ridges are destroyed by secondary tillage, the amount of soil erosion would be high.

The plot pictured in Fig. 1, bottom, has not been tilled and shows what happens when rain falls on plant residue. The surface residue do the following:

- Absorb the impact energy of the raindrops and thereby greatly reduces splash erosion.
- Provide an obstruction to the flowing water, which slows the velocity of the runoff and, therefore, the water has more time to infiltrate the soil.
- Reduce the tendency for the soil surface to seal, which would decrease infiltration.

JDPX5853

*Fig. 1 — "Rain" Being Applied to Plots Following Corn With Different Tillage Systems*

## Amount of Runoff and Erosion

In comparing the tillage systems with respect to water runoff, plots that had been moldboard plowed had the most runoff (Fig. 2). Of the first 3 inches (7.6 cm) of rain applied, 1.6 inches (4.1 cm) ran off. For the no–tillage system, 3 inches (7.6 cm) of rain resulted in 1.0 inch (2.54 cm) of runoff. The ridges formed by the chisel plow reduced the runoff to only 0.1 inch (0.25 cm). Of course, the low runoff of the chisel plowed surface would not be maintained after secondary tillage is used to level the surface and prepare a seedbed.

The most important difference between tillage systems is the amount of soil erosion caused by the runoff. Soil erosion was highest from plots that had been moldboard plowed in the fall (Fig. 3). The first 3 inches (7.6 cm) of rain carried 2.6 tons of soil per acre (5.8 t/ha). With chisel plowing and no–till, the soil erosion was less than 0.3 ton/acre (0.67 t/ha). As mentioned before with the chisel plow system, the low amount of soil erosion would not be expected after the ridges were leveled with secondary tillage and a crop planted.

The amount of water runoff following soybeans was slightly higher from moldboard plow plots compared to the water runoff from the no–till and chisel plow plots (Fig. 4). After 3 inches (7.6 cm) of rain, the runoff was 2.1 inches (5.3 cm) with moldboard plow, 1.7 inches (4.3 cm) with no–till, and 1.2 inches (3.0 cm) with chisel plow.

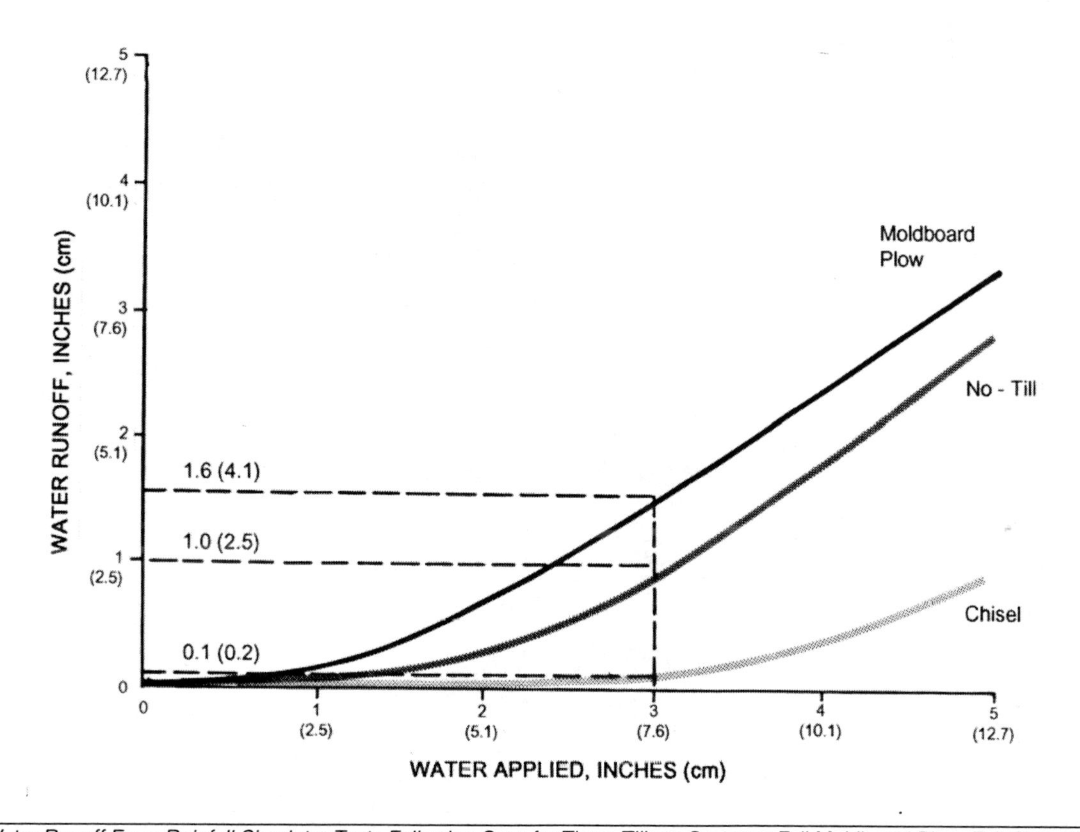


*Fig. 2 — Water Runoff From Rainfall Simulator Tests Following Corn for Three Tillage Systems: Fall Moldboard Plow, Fall Chisel Plow, and No–Till*

JDPX5855

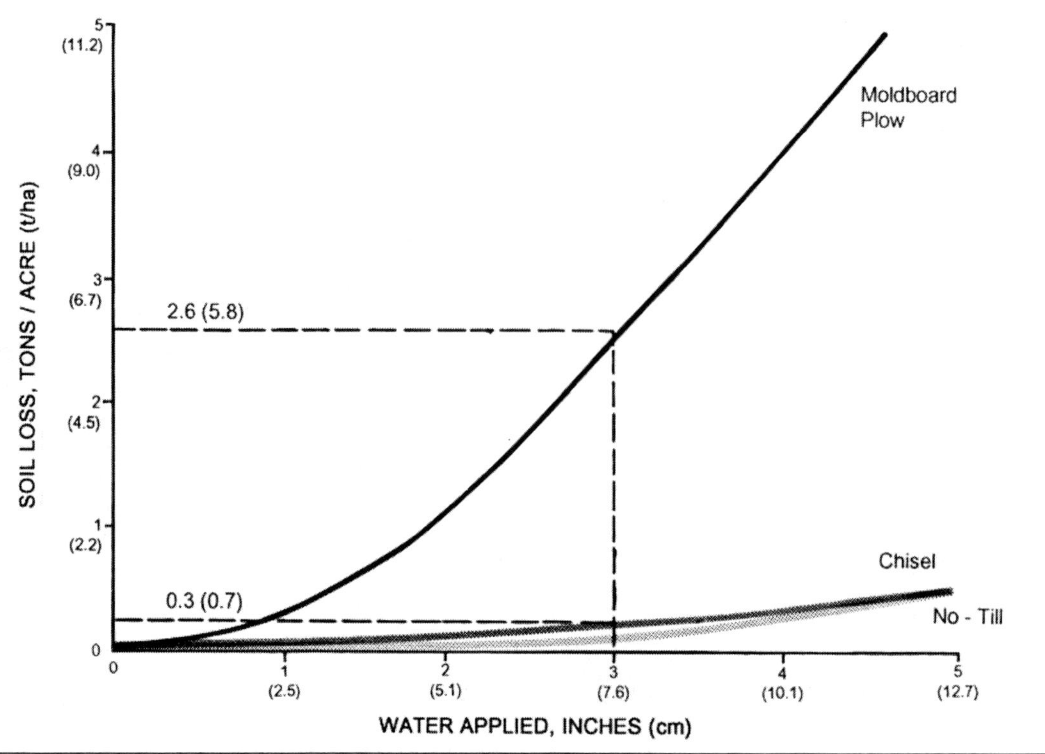


*Fig. 3 — Soil Erosion Following Corn From Rainfall Simulator Tests for Three Tillage Systems: Fall Moldboard Plow, Fall Chisel Plow, and No–Till*

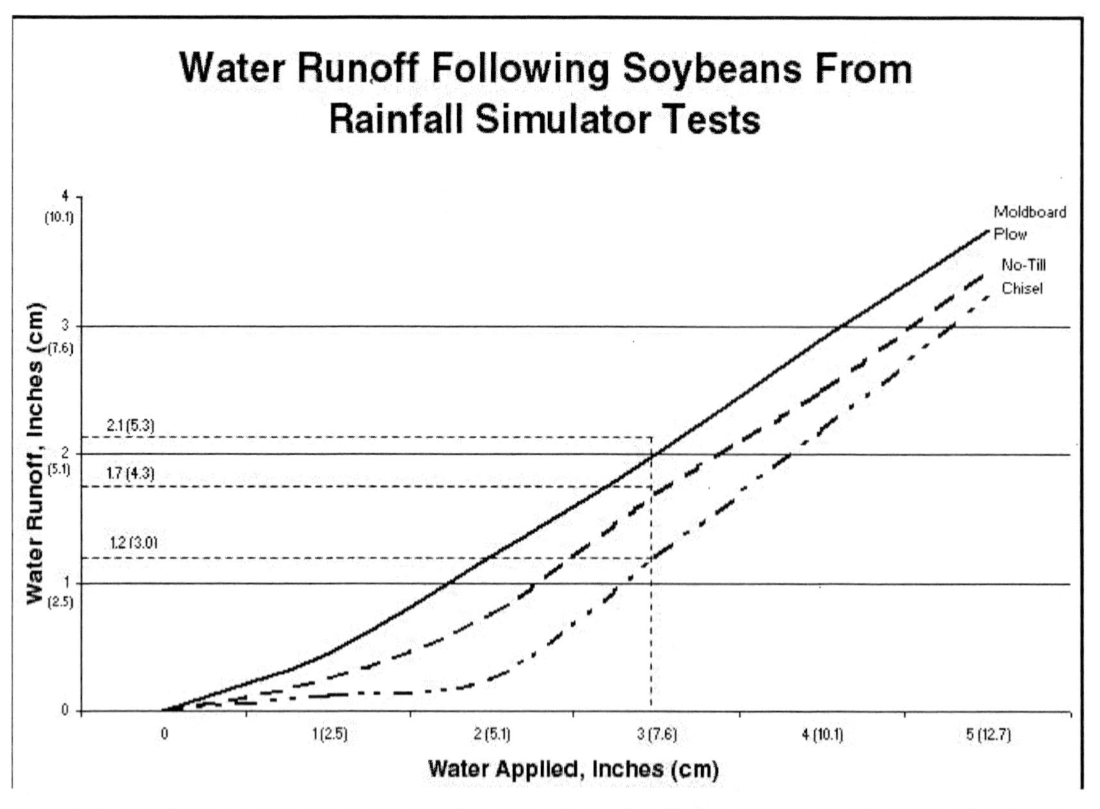


JDPX6105

*Fig. 4 — Water Runoff Following Soybeans from Rainfall Simulator Tests for Three Systems: Fall Moldboard Plow, Chisel Plow, and No–till*

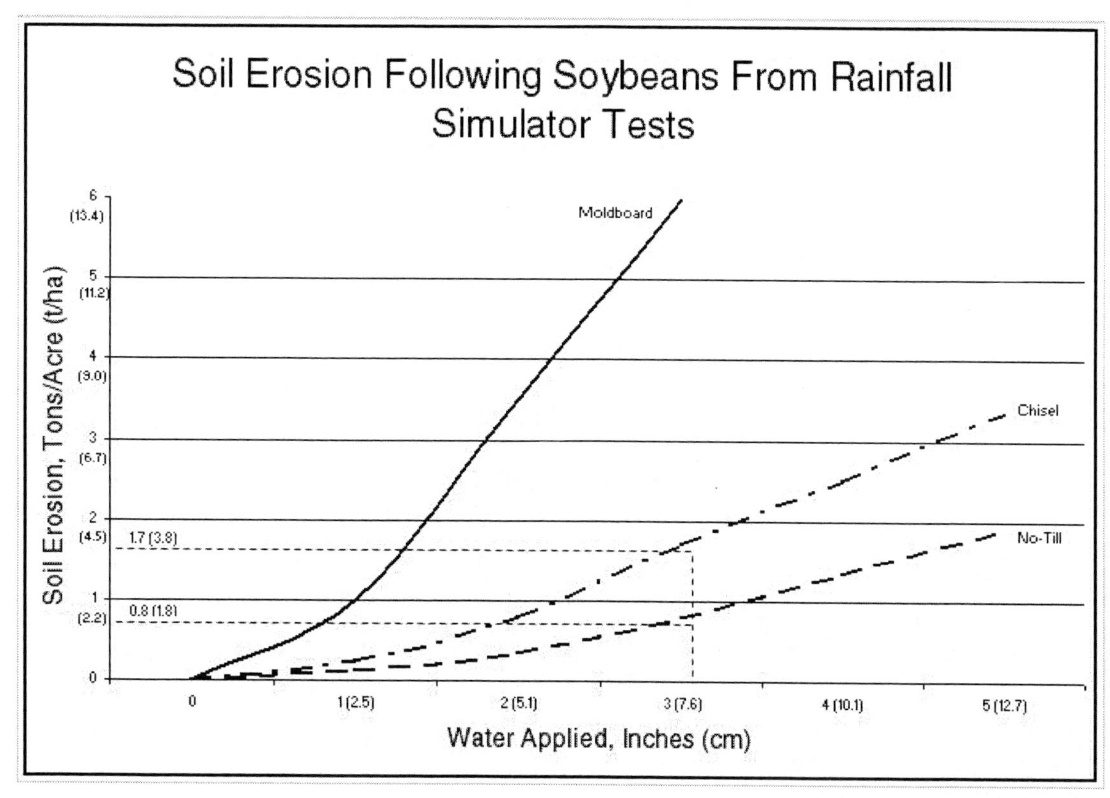


*Fig. 5 — Soil Erosion Following Soybeans From Rainfall Simulator Tests in the Spring for Three Tillage Systems: Moldboard Plow, Chisel Plow, and No–Till*

The amount of soil erosion following soybeans is much greater than following corn (Fig. 5). After the first 3 inches (7.6 cm) of rain following soybeans, the soil erosion from the moldboard plowed plots was 6 tons/acre (13.4 t/ha). From the chisel plowed plots the soil erosion was 1.7 tons/acre (3.8 t/ha), almost the same as from the moldboard plowed plots following corn. With no–till, the soil erosion was 0.8 ton/acre (1.8 t/ha) due to the first 3 inches (7.6 cm) of rain. The reasons more soil erosion occurs after soybeans than after corn are believed to be due to the following:

- Soybean residue is much more fragile than corn residue, and there is less of it. The fragile soybean residue is much easier to cover with soil, and it deteriorates faster.
- Soil following soybeans is more mellow and susceptible to soil erosion.

## Soil Organic Matter

In addition to providing a means of controlling soil erosion, plant residues are a source of organic matter. Soil organic matter is an essential part of a productive soil. It improves the physical and chemical properties of the soil and contains essential nutrients that are released upon its decomposition. Soil organic matter influences soil water–holding capacity, aggregation, tilth, water permeability, aeration, and erodibility. There are two sources of organic residues in agriculture:

- Above–surface residues
- Below–surface residues

Perennial plants like trees and shrubs add organic residues to soils primarily when leaves fall to the ground. The accumulation is on the surface. Annual species, such as grasses and most agricultural crops, add organic residues not only to the surface, but also to below the soil surface. Large amounts of residues are added below the soil surface due to the decay of roots (Fig. 6). For most crops almost equal amounts of residues are added each year in the form of roots as are added in the form of above–surface residue. Few farmers realize how much the root systems of the crops they grow aid in maintaining organic matter level in their soil.

Crop residue management pertains to what is done with or what happens to the portion of plants remaining on the soil surface after the crop is harvested.

Surface residues from plants are a tremendous natural resource. Plant residues remaining on the soil surface after harvest are not trash or waste material as they are often regarded. From a conservation perspective, crop residue management is the use of plant residues for soil improvement and protection from erosion. There are also other uses of crop residues. For example, crop residues can be collected and used for livestock feed or bedding.

Crop residue management is a system. It begins with the selection of crops to be grown and includes all aspects of how the crops are grown and harvested and how the residues are handled after harvest. Crop residue management includes all field operations performed that affect the amount of residue on the soil surface.

JDPX5856

*Fig. 6 — Roots Provide Large Amounts of Organic Matter to the Soil*

## Some History of Residue Management

When our present–day soils began to form, they had very minute quantities of organic matter. As plants developed, their roots and foliage were returned to the soil as residue. The residue decayed, and soil organic matter was formed. A decrease in soil organic matter content occurs when a virgin soil is brought under cultivation. For example, the organic matter content of three North Dakota soils decreased about 25% after 43 years of cropping.

In early practices, it was common to burn plant residues after harvesting crops like wheat or corn. Burning residue got it out of the way so it did not plug tillage equipment that was used at the time. Soils were sufficiently supplied with nutrients for the crops and crop varieties being grown. Fertility levels were improved by using crop rotations that included legume crops, like clover and alfalfa, and manures.

After World War II, inexpensive mineral fertilizers and high yielding crop varieties became available. Since the late 1950s and 1960s, effective pesticides have been developed to control weeds, insects, and diseases that were previously controlled by crop rotations and tillage. Availability and use of mineral fertilizers and pesticides helped make intensive row–crop agriculture possible. Most farmers specialized in growing two or three crops. In the Midwest for example, the major crops, by far, on most farms are now corn and soybeans, both annual crops (Fig. 7). The moldboard plow was by far the most common primary tillage tool used (Fig. 8).

JDPX6176

*Fig. 7 — Corn (Left) and Soybeans (Right) Are the Only Crops Produced by Many Farmers in the North Central Region of the U.S.*

Intensive row–crop agriculture made the soil more susceptible to erosion. The soil became more susceptible to erosion because many fields were moldboard plowed every year. The moldboard plow buried essentially all the residues, and a smooth, fine seedbed was prepared for planting. Also, with the adoption of intensive specialized agriculture, most of the windbreaks and fencerows were removed.

A widespread shift back to other crop rotations that include alfalfa or clover would result in severe economic losses to the individual farmer. Thus, there has been a need to develop crop production techniques that effectively control erosion due to wind and water without removing the economic advantages resulting from intensive production of annual crops. The need was especially true for major crops like corn, soybeans, and wheat. The most significant low–cost and widely adopted technique developed to reduce soil erosion has been conservation tillage systems.

Remember that conservation tillage systems greatly reduce the potential for soil erosion to occur. Most conservation tillage systems rely on plant residues on or near the soil surface to reduce the erosion potential. Because plant residues are so important in conservation tillage systems, the amount of plant residues produced by various crops will be covered next.

JDPX6177

*Fig. 8 — The Moldboard Plow Was the Most Common Primary Tillage Tool for Many Years*

JDPX6178

*Fig. 9 — In the Past, It Was Common to Burn Crop Residues*

## Amount of Crop Residue Produced

In the past, we were poor stewards in managing plant residue resources. The residues were commonly handled in one of the following ways:

- Burning
- Burying
- Removing

Residues were burned to get rid of them so subsequent tillage would be easier (Fig. 9). It was also believed by some that when the residues were not burned, they would absorb valuable soil moisture and nutrients during the decaying process. If residues were not burned, they were commonly buried with the moldboard plow (Fig. 8). Residues were and continue to be removed or harvested and used for livestock feed or bedding. Sometimes essentially the entire plant is harvested, like when corn is harvested for making silage (Fig. 10). Today, very few farmers burn crop residues, and use of the moldboard plow has decreased substantially in many regions. Also, the percentage of the cropland from which the entire plant is harvested is small.

The quantity of crop residues produced and remaining on farm fields after harvest is tremendous. The approximate amount of plant residue produced for selected crops is given in Fig. 11.

Plant residues are organic matter. What happens to or is done with the plant residues, therefore, influences the organic matter content of the soil. The influence that was determined in an experiment is illustrated in Fig. 12. From 0 to 7.1 tons per acre (0 to 16 t/ha) of corn or alfalfa residues were added to the soil for 11 consecutive years. Corn was grown each year. After 11 years the organic carbon content was directly related to the amount of residue added. The original organic matter content of the soil was 1.8 percent.

When all residues were removed each year, the organic matter content decreased to 1.6 percent. With 7.1 tons of residue per acre (16 t/ha) added each year, the organic matter content increased to 2.4 percent. The results of this experiment illustrate how difficult it is to make large changes in the organic matter content of the soil. It was estimated that about 4500 lb per acre (5 t/ha) of residue or about the amount produced by a corn crop yielding 80 bu per acre (2.5 t/ha) was needed to maintain the initial 1.8 percent organic carbon in the soil. Higher corn yields should maintain a slightly higher organic matter level. However, many crops do not produce as much residue as corn.

*Fig. 10 — Essentially the Entire Corn Plant Being Harvested for Silage*

| Crop | | Crop Yield | | Residue | |
|---|---|---|---|---|---|
| | | bu/acre | (t/ha) | lb/acre | (t/ha) |
| Barley | | 50 | (2.63) | 3750 | (4.21) |
| Corn | | 100 | (6.27) | 5600 | (6.27) |
| | | 150 | (9.42) | 8400 | (9.42) |
| Sorghum | | 70 | (4.47) | 3990 | (4.47) |
| Soybean | | 35 | (2.35) | 2100 | (2.35) |
| | | 50 | (3.36) | 3000 | (3.36) |
| Wheat | Winter | 30 | (2.02) | 3300 | (3.70) |
| | | 70 | (4.71) | 7700 | (8.64) |
| | Spring | 50 | (3.25) | 4750 | (5.33) |
| | | lb/acre | (t/ha) | | |
| Cotton | | 600 | (0.67) | 600 | (0.67) |
| | | 1200 | (1.34) | 1200 | (1.34) |
| Rice | | 5500 | (6.16) | 8250 | (9.24) |

*Fig. 11 — Approximate Amount of Residue Produced by Selected Crops*

## Determining Percent Residue Cover

Residue management techniques that include conservation tillage systems are widely promoted as the best low–cost means of reducing erosion of agricultural land. Many farmers have chosen conservation tillage as part of their conservation compliance plans to meet requirements of the 1985 Food Security Act. Conservation tillage means that a minimum of 30 percent of the soil surface must be covered with residues—measured after planting. Some of the compliance plans specify even greater percentages of residue cover. A relatively new term, residue management refers to systems with residues higher than with conventional tillage systems. Even if compliance is not a concern, farmers need to know the amount of residue cover in their fields to better control soil erosion. Thus, determining residue cover is critical for proper residue management.

There are several methods to measure or estimate residue cover including:

- Line–transect
- Photo comparison
- Prediction

The line–transect and photo comparison methods can be used to actually measure the amount of the soil surface covered with residue in a field. The predicition method is used when it is desired to estimate the amount of residue that will be on the soil surface after one or more field operations.

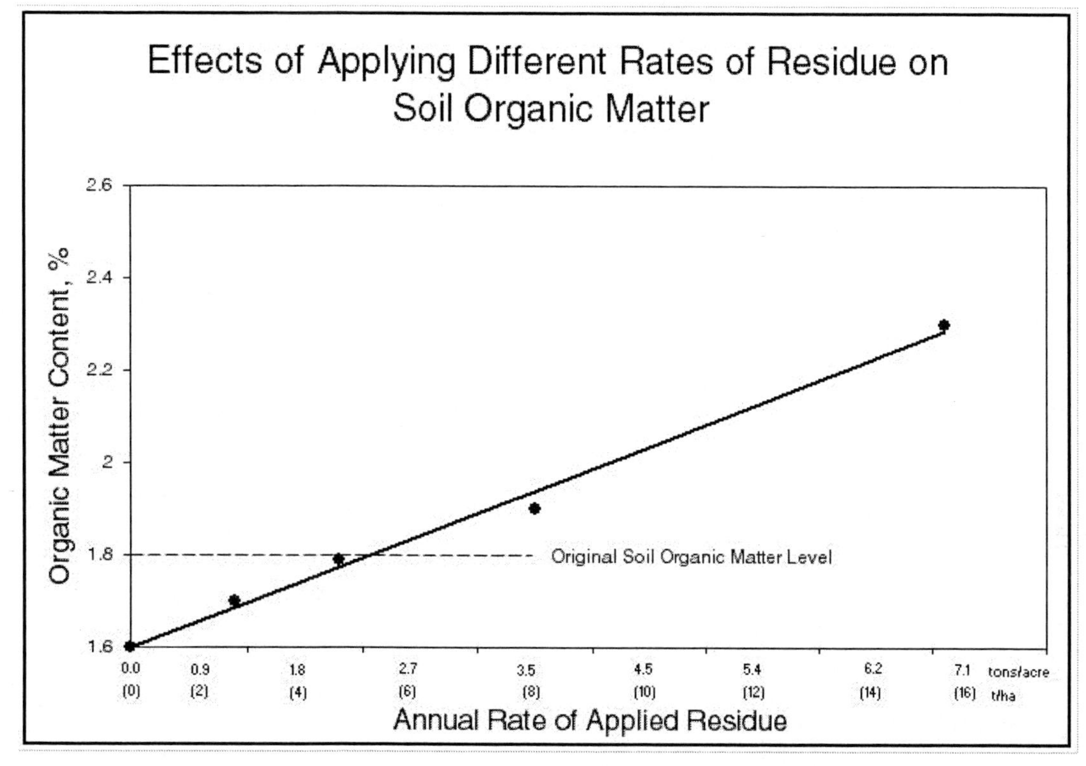


JDPX6107

*Fig. 12 — Effects of Applying Different Rates of Residue for 11 Consecutive Years on Soil Organic Matter*

## Line–Transect Method

The line–transect is an easy and accurate way of measuring residue cover. Get a light rope, about 70 feet (20 m) long. Tie 100 knots in the rope about 6 inches (15 cm) apart. Instead of a rope, you can use a measuring tape as long as it has about 100 equally spaced points (Fig. 13). The following is a step–by–step procedure for using the line transect method.

Step 1. Select an area in the field that is representative of the entire field.

Step 2. Stretch the rope or tape diagonally across the crop rows so that it crosses at least one width of the farming implements used (Fig. 14).

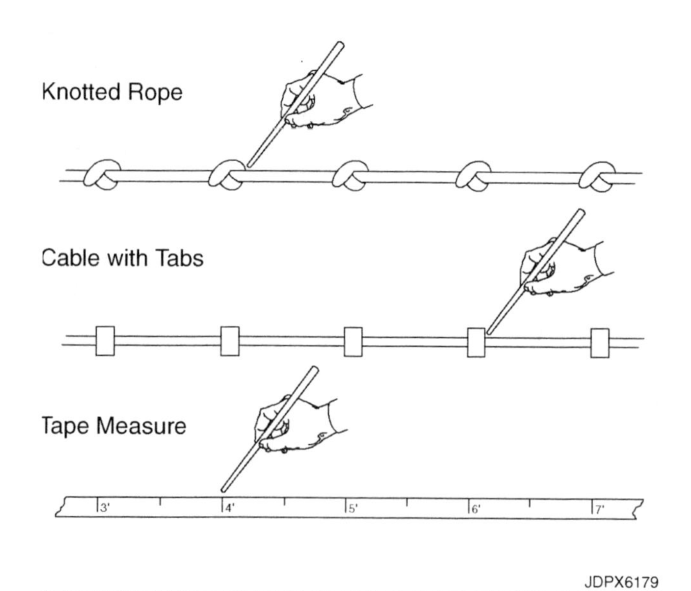


*Fig. 13 — Common Types of Lines Used to Measure Residue Transect*

Step 3. Count the number of knots in the rope that is directly over a piece of residue. For accurate measurement several rules must be followed when counting:

- Do not move the tape while counting.
- Look straight down at the same edge or corner of each knot.
- Leaning from side to side will cause you to overestimate the residue cover.
- Count only those knots or marks on a tape that have residue directly under the corner or edge you look at (Fig. 15).
- Count only residue that is large enough to dissipate the energy of a rain drop that occurs during an intense storm.

Residue about 3/32 inch (25 mm) in diameter is the minimum size suggested for a piece of residue to be counted.

Step 4. The percent residue cover is equal to the number of points counted.

Step 5. Repeat steps 1 to 4 (at least three times) in the field and average the counts to get an accurate measurement.

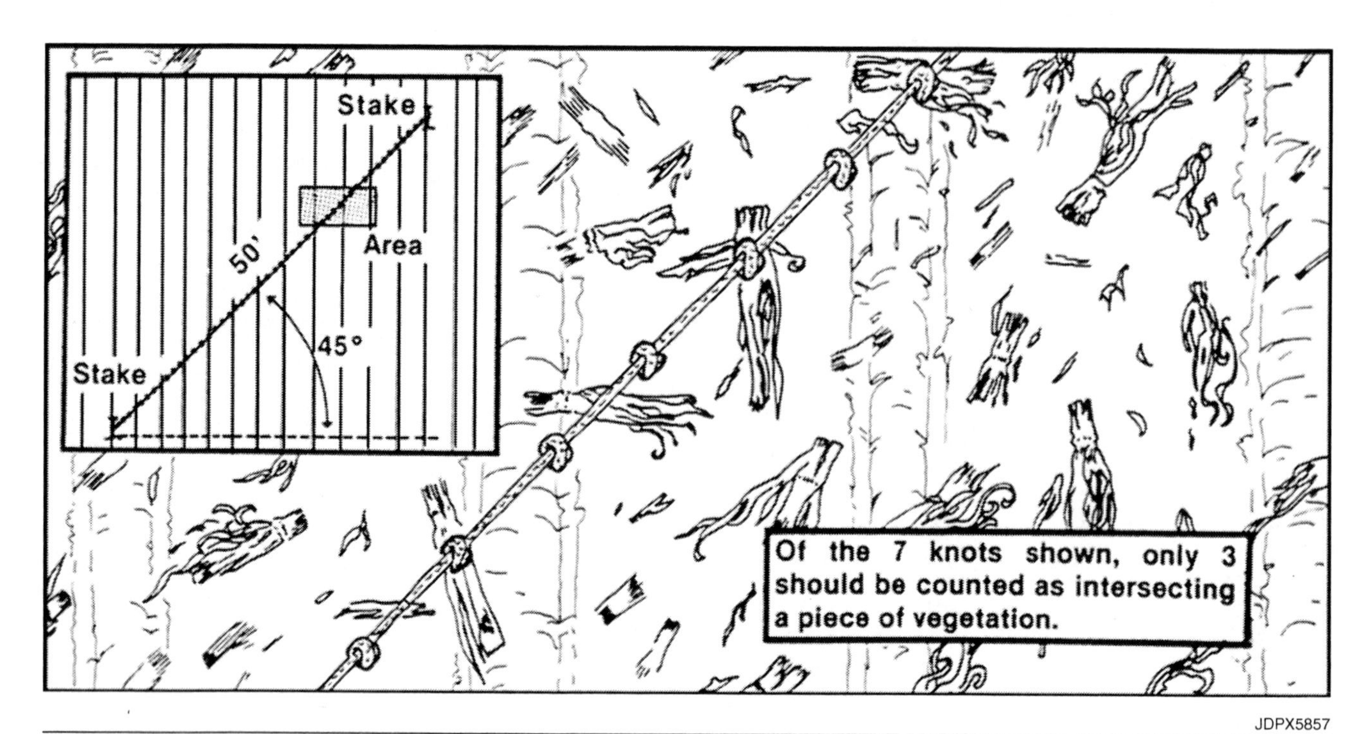


*Fig. 14 — Overview (Insert) and Close–Up of the Line–Transect Method of Measuring Residue Cover*

## Photo Comparison Method

Residue cover can be roughly estimated by comparing photographs of known residue cover percentages with actual field conditions. Some photographs of different levels of residue cover are shown in Fig. 15 for corn, and Fig. 16 for soybeans.

To use the photo–comparison method, stand in the field, look straight at the soil surface and compare the observed residue cover with that in the photographs. This method is not as accurate as the line–transect method.

## Prediction Method

The prediction method can be used to predict the amount of residue that you will have in a field using a specific tillage system. The method is based on the fact that a field operation (tillage, planting, etc.) leaves only a percentage of the residues on the soil surface that was there before the operation.

To use the prediction method you must:

- Determine whether the residue in the field is non–fragile or fragile.
- Estimate the percentage of the residue cover remaining on the soil surface after each field operation to be used. It may also be necessary to take into consideration the amount of residue decomposition that occurs over winter.

### Non–Fragile or Fragile Crop Residues

Various crops are listed in Fig. 17 according to type of residue produced, non–fragile or fragile. This list is a subjective classification based on the nature of the residue in regard to:

- Ease of decomposition by weather
- Ease of residue cover by tillage operations

Characteristics of plant residue such as composition and size of leaves and stems, density and quantity of the residue produced were considered in deciding whether the residue of a specific crop is non–fragile or fragile. The residue of a crop like soybeans is considered to be fragile because essentially all the residues are damaged in passing through the combine, the stems are small in diameter, and the leaves are small in size and fall from the plant well before harvest.

Also, many fragile residues have high nitrogen content, which favors microbial breakdown. In contrast, corn residues are classified as non–fragile. Corn stalks, leaves, and cobs are individually large in size and quite durable to the elements, and the total mass of residues produced is great.

### Residue Cover Remaining

Once you have determined whether the type of residue you have is non–fragile or fragile, you are ready to estimate the percentage of soil surface that will be covered after one or more field operations.

Many factors affect the percentage of the soil surface covered with residues after a pass with a tractor and tillage or planting equipment. In addition to the type of residue, the main factors are the following:

- Percentage of the soil surface covered with residues before the operation
- Speed and depth of the operation
- Condition of residue (wet or dry)
- Condition of soil (wet or dry)
- Attachments on the tillage and planting equipment
- Residue weathering or resistance to cutting

Each tillage or planting operation leaves a percent of the residue that was present just prior to that operation. The numbers in Fig. 18 are an estimate of these remaining percentages.

### Predicting Residue Cover for a Tillage System

The residue cover remaining after planting for a selected tillage and planting system can be estimated using the type of residue given in Fig. 17 and the residue cover remaining values given in Fig. 18. The following example illustrates how to use the values. Assume a tillage and planting system used on a field of corn residue consists of four operations: chisel plowing in the fall and in the spring, disking, field cultivating, and planting. From Fig. 17 corn residue is considered to be non–fragile. The four operations and the percentage residue cover remaining from Fig. 18 are listed below:

1. Fall disk–chisel plow with twisted points (60%)
2. Tandem disk, equipped with finishing disk with 7.5 inch (19 cm) blade spacing (55%)
3. Field cultivate with sweeps 10 inches (25.4 cm) wide (75%)
4. Plant with conventional planter equipped with double disk openers (90%)

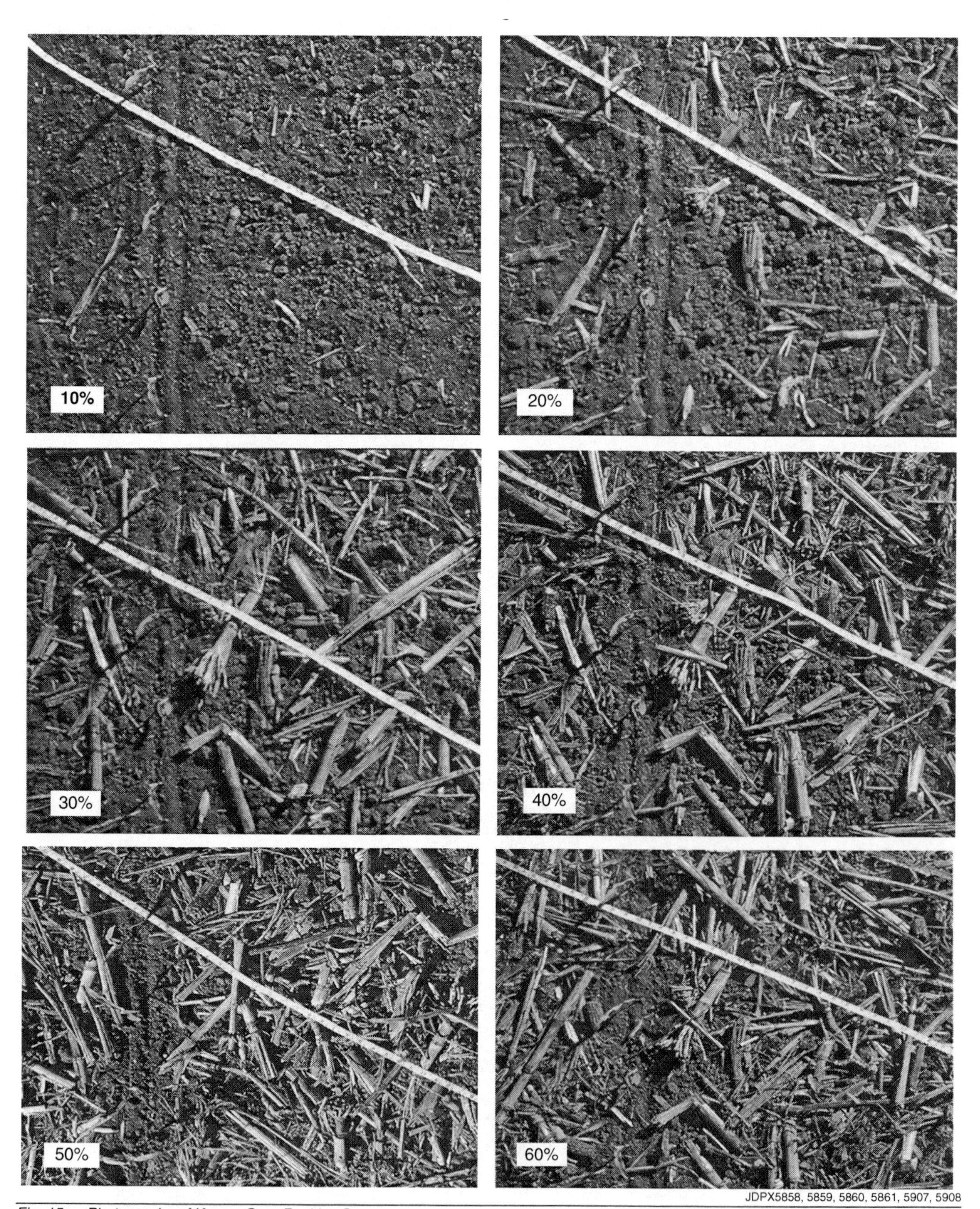


JDPX5858, 5859, 5860, 5861, 5907, 5908

*Fig. 15 — Photographs of Known Corn Residue Cover*

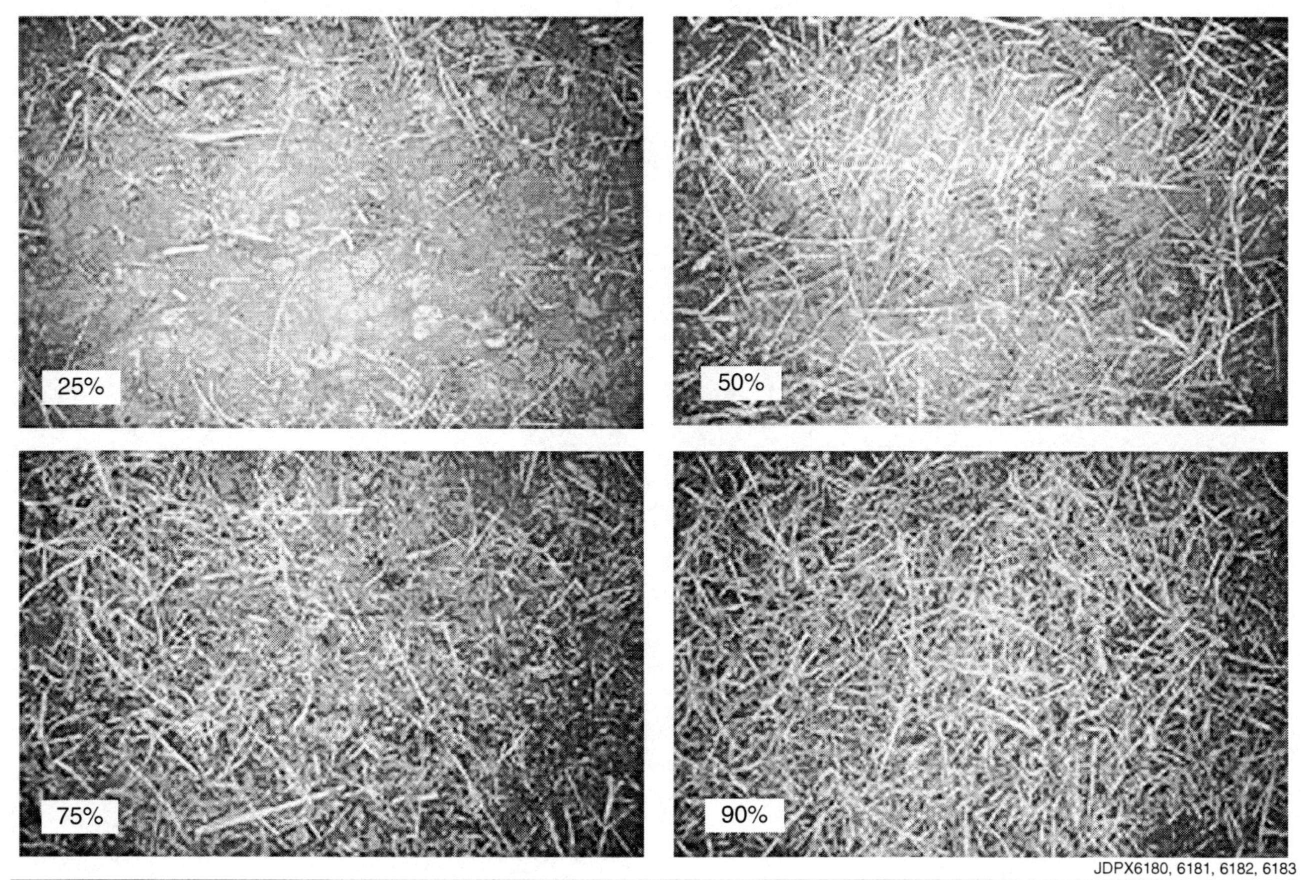


*Fig. 16 — Photographs of Known Soybean Residue Cover*

| Non–Fragile | Fragile | Non–Fragile | Fragile |
|---|---|---|---|
| Alfalfa or Legume hay | Canola/Rapeseed | Pasture | Mint |
| Barley[a] | Cotton | Pineapple | Mustard |
| Buckwheat | Dry beans | Popcorn | Peanuts |
| Corn | Dry peas | Rice | Potatoes |
| Flaxweed | Fall seeded cover crops | Rye[a] | Safflower |
| Forage seed | Flower seed | Sorghum | Soybeans |
| Forage silage | Grapes | Sugarcane | Sunflowers |
| Grass hay | Green peas | Tobacco | Sweet Potatoes |
| Millet | Guar | Tritacale[a] | Vegetables |
| Oats[a] | Lentils | Wheat[a] | |

*a. If a combine is equipped with a straw chopper or the straw is otherwise cut into small pieces, small grain residue should be considered as being fragile. If the yield of a non–fragile crop is low, it should be considered as fragile.*

*Fig. 17 — Types of Residue Produced by Various Crops*

To leave more residue on the soil surface would require that one or more of the tillage operations not be performed or that attachments be used which do not cover as much of the residue. For example, sweeps could be used on the chisel plow instead of twisted points. Of course, the most residue would be left on the soil surface if the no–tillage system were used. Consider the prediction method as only a rough estimate because many variables are involved. For a specific implement, the type and condition of the residue and soil, and the adjustments, speed, and depth of the implement will all affect the percent residue remaining.

Actual residue cover should be determined by in–field measurements, such as with the line–transect method. The values in Fig. 18 are useful for planning tillage operations. The values can be used to obtain a general idea of how much residue will remain after a sequence of tillage and planting operations.

Assume an initial residue cover of 95% and that the residue remaining after winter weathering is 90% of the surface residue percentage before winter. The predicted residue cover after planting is calculated as follows:

- Cover after harvest 95%
- Cover after chisel plowing = 95% x 0.60 = 57%
- Cover after winter weathering = 57% x 0.90 = 51%
- Cover after tandem disking 51% x 0.55 = 28%
- Cover after field cultivating = 28% x 0.75 = 21%
- Cover after planting –22 21% x 0.90 = 19% For the above tillage and planting system, the residue cover remaining on the soil surface would be reduced to 19%. The 19% cover is less than the 30% needed for conservation tillage.

| | Type of Residue | |
|---|---|---|
| | **Non–Fragile** | **Fragile** |
| | **Percent of Residue Remaining** | |
| CLIMATE EFFECTS: | | |
| Over winter weathering: | | |
| Following summer harvest | 70–80 | 65–85 |
| Following fall harvest | 80–95 | 70–80 |
| In northern climates with long periods of snow cover and frozen conditions, weathering may reduce residue levels only slightly, while in warmer climates, weathering losses may reduce residue levels significantly. | | |
| Moldboard plow | 0–10 | 0–5 |
| Moldboard plow–uphill furrow (Pacific Northwest Region only) | 30–40 | |
| Disk plow | 10–20 | 5–15 |
| SUBSOILERS: | | |
| Paratill/Paraplow | 80–90 | 75–85 |
| "V" ripper/subsoiler 12–14 in. (30–36 cm) deep 20 in. (50 cm) spacing | 70–90 | 60–80 |
| Combination tools: | | |
| Subsoiler–chisel | 50–70 | 40–50 |
| Disk–subsoiler | 30–50 | 10–20 |
| CHISEL PLOWS with: | | |
| Sweeps | 70–85 | 50–60 |
| Straight spike points | 60–80 | 40–60 |
| Twisted points or shovels | 40–60 | 20–30 |
| COMBINATION CHISEL PLOWS: | | |
| Coulter–chisel plow with: | | |
| Sweeps | 60–80 | 40–50 |
| Straight spike points | 50–70 | 30–40 |
| Twisted points or shovels | 40–60 | 20–30 |
| Disk–chisel plow with: | | |
| Sweeps | 60–70 | 30–50 |
| Straight spike points | 50–60 | 30–40 |
| Twisted points or shovels | 30–50 | 20–30 |
| UNDERCUTTERS: | | |
| Stubble–mulch sweep or blade plow with: | | |
| Sweeps or "V"–Blades > 30 in. (76 cm) wide | 85–95 | 70–80 |
| Sweeps 20–30 in. (50 cm) wide | 80–90 | 65–75 |
| DISKS: | | |
| Offset: Heavy plowing > 10 in. (25 cm) spacing | 25–50 | 10–25 |
| Primary cutting > 9 in. (23 cm) spacing | 30–60 | 20–40 |
| Finishing 7–9 in. (18–23 cm) spacing | 40–70 | 25–40 |
| Tandem: Heavy plowing > 10" (25 cm) spacing | 25–50 | 10–25 |

| | Type of Residue | |
|---|---|---|
| | Non–Fragile | Fragile |
| | Percent of Residue Remaining | |
| Primary cutting > in. (23 cm) spacing | 30—60 | 20–40 |
| Finishing 7–9 in. (18–23 cm) spacing | 40–70 | 25–40 |
| Light tandem disking after harvest, before other tillage | 70–80 | 40–50 |
| One–way disk with: | | |
| 12–16 in. (30–40 cm) blades | 40–50 | 20–40 |
| 18–30 in. (46–76 cm) blades | 20–40 | 10–30 |
| FIELD CULTIVATORS: (including leveling attachments) | | |
| Field cultivator as primary tillage operation: | | |
| Sweeps 12–20 in. (30–50 cm) | 60–80 | 55–75 |
| Sweeps or shovels 6–12 in. (15–30 cm) | 35–75 | 50–70 |
| Duckfoot points | 35–60 | 30–55 |
| Field cultivator as secondary tillage operation: | | |
| Sweeps 12–20 in. (30–50 cm) | 80–90 | 60–75 |
| Sweeps or shovels 6–12" (15–30 cm) | 70–80 | 50–60 |
| Duckfoot points | 60–70 | 35–50 |
| FINISHING TOOLS: | | |
| Combination finishing tool with: | | |
| Disks, shanks, and leveling attachment | 50–70 | 30–50 |
| Spring teeth and rolling baskets | 70–90 | 50–70 |
| Harrows: | | |
| Springtooth (coil tine) | 60–80 | 50–70 |
| Spike tooth | 70–90 | 60–80 |
| Flex–tine tooth | 75–90 | 70–85 |
| Roller harrow (cultipacker) | 60–80 | 50–70 |
| Packer roller | 90–95 | 90–95 |
| Rotary Tillers: | | |
| Secondary operation 3 in. (7.6 cm) deep | 40–60 | 20–40 |
| Primary operation 6 in. (15 cm) deep | 15–35 | 5–15 |
| RODWEEDER: | | |
| Plain rotary rod | 80–90 | 50–60 |
| Rotary rod with semi–chisels or shovels | 70–80 | 60–70 |
| STRIP TILLAGE MACHINES: | | |
| Rotary tiller, 12 in. (30 cm) wide strips tilled on 40 in. (102 cm) rows | 60–75 | 50–60 |
| ROW CULTIVATORS: 30 in. (76 cm) WIDE ROWS OR WIDER | | |
| Single sweep per row | 75–90 | 55–70 |
| Multiple sweeps per row | 75–85 | 55–65 |
| Finger wheel cultivator | 65–75 | 50–60 |
| Rolling disk cultivator | 45–55 | 40–50 |
| Ridge–till cultivator | 20–40 | 5–25 |
| UNCLASSIFIED MACHINES: | | |
| Anhydrous ammonia applicator | 75–85 | 45–70 |
| Anhydrous ammonia applicator with closing disks | 60–75 | 30–50 |

| | Type of Residue | |
|---|---|---|
| | Non–Fragile | Fragile |
| | Percent of Residue Remaining | |
| Subsurface manure applicator | 60–80 | 40–60 |
| Rotary hoe | 85–90 | 80–90 |
| Bedders, listers, and hippers | 15–30 | 5–20 |
| Furrow diker | 85–95 | 75–85 |
| Mulch treader | 70–85 | 60–75 |
| DRILLS: | | |
| Hoe opener drills | 50–80 | 40–60 |
| Semi–deep furrow drill or press drill 7–12 in. (18–30 cm) | 70–90 | 50–80 |
| Deep furrow drill with > 12 in. (30 cm) spacing | 60–80 | 50–80 |
| Single disk opener drills | 85–100 | 75–85 |
| Double disk opener drills (conventional) | 80–100 | 60–80 |
| No–till drills and drills with the following attachments in standing stubble: | | |
| Smooth no–till coulters | 65–85 | 70–85 |
| Ripple or bubble coulters | 80–85 | 65–85 |
| Fluted coulters | 55–70 | 40–60 |
| No–till drills and drills with the following attachments in flat residues: | | |
| Smooth no–till coulters | 65–85 | 50–70 |
| Ripple or bubble coulters | 60–75 | 45–65 |
| Fluted coulters | 55–70 | 40–60 |
| Air seeders: (Refer to appropriate field cultivator or chisel plow depending on the type of ground engaging devise used.) | | |
| Air drills: (Refer to corresponding type of drill opener.) | | |
| ROW PLANTERS: | | |
| Conventional planters with: | | |
| Runner openers | 85–95 | 80–90 |
| Staggered double disk openers | 90–95 | 85–95 |
| Double disk openers | 85–95 | 75–85 |
| No–till planter with: | | |
| Smooth coulters | 85–95 | 75–90 |
| Ripple coulters | 75–90 | 70–85 |
| Fluted coulters | 65–85 | 55–80 |
| Strip–till planters with: | | |
| 2 or 3 fluted coulters | 60–80 | 50–75 |
| Row cleaning devices, 8–14 in. (20–36 cm) wide bare strip using brushes, spikes, furrowing disks, or sweeps | 60–80 | 50–60 |
| Ridge–till planter | 40–60 | 20–40 |

*Fig. 18 — Percent Residue Cover Remaining on the Soil Surface After Weathering or Specific Field Operations*

## Summary

Soil organic matter is an essential part of a productive soil. Two sources of organic residues from plants are:

- Above ground residues
- Below ground residues

Crop residue management pertains to what is done with or what happens to the portion of plants remaining on the soil surface after the crop is harvested. It includes all field operations performed that affect the amount of residue on the soil surface. The most widely adopted residue management technique is conservation tillage. The major reasons for the interest in adopting conservation tillage are to reduce:

- Soil erosion due to wind and water
- Machinery costs
- Time and labor costs

Conservation tillage is any system that provides resistance to the erosive effects of wind, rain, and flowing water. Conservation tillage is often defined as any crop production system that provides one of the following:

- A residue cover of at least 30 percent after planting to reduce soil erosion due to water
- At least 1000 pounds per acre (1121 kg/ha) of flat, small grain residue, or the equivalent, on the soil surface during the critical erosion period to reduce soil erosion due to wind

Percent residue cover can be measured or estimated with one of the following methods:

- Line–transect
- Photo comparison
- Prediction

Plant residues on the soil surface are very effective in reducing soil erosion due to wind and water. In rainfall simulator tests conducted by the University of Illinois, soil erosion caused by 3 inches (7.6 cm) of rain was 2.6 tons of soil per acre (5.8 t/ha) from moldboard plowed plots. With chisel plowing and no–till, the soil erosion was less than 0.3 ton/acre (0.67 t/ha). Following soybeans, the soil erosion caused by 3 inches (7.6 cm) of rain from moldboard plowed plots was 6 tons/acre (13.4 t/ha), from chisel plowed plots 1.7 tons/acre (3.8 t/ha), and with no–till 0.8 tons/acre (1.8 t/ha).

## Test Yourself

### Questions

1. Define crop residue management.
2. (Fill in blanks.) Two sources of soil organic matter are __________ and ____________.
3. List three methods for measuring or estimating percent residue cover.
4. (T/F) The 1985 Food Security Act requires that 45 percent of the soil surface must be covered with residue in a conservation tillage system.
5. What determines whether residue is fragile or non–fragile? List three examples of each type.
6. (T/F) V–ripper/subsoilers leave more non–fragile residue on the soil surface than any other commonly used tillage implement.
7. (Fill in blanks.) Major reasons for increased adoption of conservation tillage are _________ and _________.
8. Define conservation tillage.
9. List three ways by which surface residues reduce soil erosion.
10. (Select one.) Which system would likely cause the greatest amount of water runoff and soil erosion: moldboard plowing in the fall, chisel plowing in the fall, no–till?

# Tractor and Implement Preparation and Adjustment

4

## Introduction

Both the tractor and the tillage tool that it pulls must be properly set up for maximum tillage efficiency and desired performance. Adjustments to attain these objectives are related primarily to traction, flotation, and soil compaction. These factors are closely related, and changes in soil–surface conditions, soil contact area, and vehicle weight will directly affect all three.

Traction is the linear force, pull or draft, resulting from torque applied to tractor tires. Traction is usually thought to be good when wheel slippage is not excessive.

Flotation is the ability of tires to stay on top of the soil surface or resist sinking into the soil.

Both traction and flotation are directly related to vehicle weight, soil conditions, and contact area between the tire and soil surface.

Compaction is the packing or firming of soil caused by wheel traffic. It is usually undesirable because it can restrict movement of air, water, and crop roots in the soil.

Several means are available to increase traction and flotation while reducing the potential for soil compaction. Tractor adjustments for traction and flotation and general guidelines for implements will be discussed in this chapter. For a detailed discussion of soil compaction and its management, see Chapter 2.

## Tractor Setup

Tractors are built with ample size and power for specific farming operations. But, to get maximum benefit from that power, additional weight may be required (Fig. 1) to gain maximum drawbar pull and sufficient traction for high draft tillage implements.

Adding ballast (weight) to drive wheels and tractor front end is the most common method of improving traction and helping increase drawbar pull.

Front–end weight (Fig. 1) may also be required for tractor stability, particularly with integral and heavy–draft, semi–integral equipment.

A tractor will provide best performance when it is equipped with tires that are large enough relative to tractor weight to give a soft ride (low stiffness). In most situations, tires should be inflated to operate at rated deflections, i.e., minimum pressures to support static weight carried on each axle. The larger the tire air volume the better. For radial tires, a rule of thumb is that the tires should require no more than 14 psi (0.97 bar), preferably less, to support the ballasted static axle load (Fig. 1). Best results will be obtained when there is more tire on the tractor than might have been thought necessary. A tractor being set up for heavy tillage work should be considered in terms of bigger tires and moderate to light weight rather than smaller tires and heavy weight.

Three major items to be considered are:

- Total tractor weight and static weight split (weight distribution on front and rear axles)
- Type of ballast used (cast and/or liquid weight)
- Tire inflation pressures

None of these can be considered independent of the others to achieve optimum tractor performance. In addition to improved traction (reduced slippage and higher fuel efficiency) and improved flotation, a properly set-up tractor will provide the following benefits:

1. Reduced compaction
2. Improved ride
3. Reduced tire wear
4. Improved sidehill stability
5. Better control of power hop on mechanical front wheel drive (MFWD) and four wheel-drive (4WD) tractors.

## Static Weight Split and Ballast

Keep in mind that draft for the same operation is usually much less in sand than in dry clay, but that any extra weight increases rolling resistance in sand or other loose soil. So, to be effective, ballast must be adjusted for soil type and conditions as well as implement load.

## Two-Wheel Drive Tractors

Two–wheel drive tractors will achieve optimum operating efficiency when static tractor weight split is set at approximately 25–35% on the front axle and 75–65% on the rear axle. The upper end of the range on the front axle should be used when heavy rear hitch mounted implements are used. To obtain correct static tractor weight split, cast weights are recommended. Liquid ballast of up to 75% fill can be used in the rear tires, but ride comfort is best if cast weight or a combination of cast and liquid weight is used.

*Fig. 1 — Weight Transfer Improves Traction—Part of Front–end Weight Is Transferred to Rear Wheels; Part Balances Rear Loads*

**IMPORTANT: Liquid ballast has an extreme stiffening effect at low inflation pressures. A partial liquid fill, used in combination with cast weights, will provide the best ride. Do not exceed 75% liquid fill (valve–stem level).**

## MFWD Tractors

**MFWD tractors** should have a static tractor weight split of 35–40% on the front and 65–60% on the rear axle. To achieve this, use cast weights on the rear and a combination of liquid and cast weights in the front.

**IMPORTANT: Liquid ballast has an undesirable stiffening effect at low inflation pressures; therefore, liquid ballast in the rear tires should be avoided when a power hop condition exists. If rear liquid weight is used, do not exceed 40% fill (valve stem at four o'clock position) and preferably not more than 25–33% fill. If front liquid weight is used, do not exceed 75% fill (valve stem level). The liquid fill must be the same in all tires on the same axle.**

### Four–Wheel Drive Tractors

Four–wheel drive tractors should have the following static tractor weight split:

- 51–55% on the front axle and 49–45% on the rear axle for standard towed implements.
- 55–60% on the front axle and 45–40% on the rear axle for hitch mounted implements.
- 55–65% on the front axle and 45–35% on the rear axle for towed implements causing high down loads on draw-bars.

To obtain the recommended static weight split, use a combination of liquid and cast weights in the rear and cast weights on the front.

**CAUTION: Installing liquid ballast in tires of any tractor requires special equipment and training. Have the job done by a tractor dealer or tire service store.**

**IMPORTANT: Liquid ballast has an undesirable stiffening effect at low pressures; therefore, liquid ballast in the rear tires should not exceed 40% fill (valve stem at four o'clock position) and preferably not more than 25–33% fill. The liquid fill must be the same in all rear tires. Added ballast is not normally required on the front axle; however, if it is required, cast weights are preferred.**

Verify the correct static tractor weight split by weighing each tractor axle or by consulting the tractor dealer.

## Increased Speed

Adding weight is not the only way to improve traction. Pulling a lighter load at a higher speed reduces wheel–slip without increasing soil compaction. For example, reducing plow size by one bottom may permit shifting up one or two gears. But this is not recommended if frequent soil obstructions are encountered. High-speed operation may also require switching to an implement such as a field cultivator instead of a disk to avoid excessive ridging.

### Inflation Pressure Adjustments

Recommended tire inflation pressure depends on static load and the number of tires mounted on the axle. Maintain the Rated Deflection determined by the tire manufacturer to operate the tire at optimum traction. Inflation pressures of a large radial tire should be set as shown in Table 1. To use the table, determine the static weight on each axle of the tractor, then each tire on the axle. Find that weight for the tire size and configuration and read the recommended minimum pressure. For radial front tires and all bias tires, use information provided by the tire manufacturer for recommended inflation pressures.

**IMPORTANT: All tires on the same axle should be inflated to the same pressure.**

## TIRE LOAD LIMITS (LB AT VARIOUS COLD INFLATION PRESSURES IN PSI)

**Conventional Size Radial Drive Wheel Tires**

| | | Tire Inflation (PSI) | | | | | | | | | | | | | | |
|---|---|---|---|---|---|---|---|---|---|---|---|---|---|---|---|---|
| Tire Size | Tire Config. | 6 | 7 | 8 | 9 | 10 | 12 | 14 | 16 | 18 | 20 | 22 | 24 | 26 | 28 | 30 |
| 18.4R38 | Singles | NR | NR | 3520 | 3740 | 3960 | 4440 | 4800 | 5200 | 5880 | 6000 | 6400 | 6600 | | | |
| | Duals | 2640 | 2820 | 3100 | 3290 | 3510 | 3910 | 4280 | 4630 | 5000 | 5260 | 5590 | 5810 | | | |
| | Triples | 2460 | 2620 | 2890 | 3070 | 3270 | 3640 | 3990 | 4310 | 4660 | 4900 | 5210 | 5410 | | | |
| 18.4R42 | Singles | NR | NR | 3740 | 3960 | 4180 | 4680 | 5080 | 5520 | 6000 | 6400 | 6600 | 6950 | | | |
| | Duals | 2710 | 3010 | 3290 | 3480 | 3680 | 4120 | 4470 | 4860 | 5280 | 5630 | 5810 | 6120 | | | |
| | Triples | 2530 | 2800 | 3070 | 3250 | 3430 | 3840 | 4170 | 4530 | 4920 | 5250 | 5410 | 5700 | | | |
| 18.4R46 | Singles | NR | NR | 3860 | 4180 | 4400 | 4940 | 5360 | 5840 | 6150 | 6600 | 6950 | 7400 | 7850 | 8060 | 8550 |
| | Duals | 2900 | 3200 | 3400 | 3680 | 3870 | 4350 | 4720 | 5140 | 5410 | 5810 | 6120 | 6510 | 6910 | 7080 | 7520 |
| | Triples | 2710 | 2980 | 3170 | 3430 | 3610 | 4050 | 4400 | 4790 | 5040 | 5410 | 5700 | 6070 | 6440 | 6600 | 7020 |
| 20.8R38 | Singles | NR | NR | 4300 | 4540 | 4800 | 5360 | 5840 | 6400 | 6800 | 7150 | 7600 | 8050 | | | |
| | Duals | 3200 | 3480 | 3780 | 4000 | 4220 | 4720 | 5140 | 5630 | 5980 | 6290 | 6690 | 7080 | | | |
| | Triples | 2980 | 3250 | 3530 | 3720 | 3940 | 4400 | 4790 | 5250 | 5580 | 5860 | 6230 | 6600 | | | |
| 20.8R42 | Singles | NR | NR | 4540 | 4800 | 5080 | 5680 | 6150 | 6800 | 7150 | 7600 | 8050 | 8550 | | | |
| | Duals | 3290 | 3680 | 4000 | 4220 | 4470 | 5000 | 5410 | 5980 | 6290 | 6690 | 7080 | 7520 | | | |
| | Triples | 3070 | 3430 | 3720 | 3940 | 4190 | 4660 | 5040 | 5580 | 5860 | 6230 | 6600 | 7010 | | | |
| 24.5R32 | Singles | NR | NR | 5080 | 5520 | 5840 | 6400 | 7150 | 7600 | 8250 | 8800 | 9100 | 9650 | | | |
| | Duals | 3780 | 4120 | 4470 | 4860 | 5140 | 5630 | 6290 | 6690 | 7260 | 7740 | 8010 | 8490 | | | |
| | Triples | 3530 | 3840 | 4170 | 4530 | 4790 | 5250 | 5860 | 6230 | 6770 | 7220 | 7460 | 7910 | | | |
| 30.5LR32 | Singles | NR | NR | 6150 | 6600 | 6950 | 7600 | 8550 | 9100 | 9650 | | | | | | |
| | Duals | 4470 | 5000 | 5410 | 5810 | 6120 | 6690 | 7520 | 8010 | 8490 | | | | | | |
| | Triples | 4170 | 4660 | 5040 | 5410 | 5700 | 6230 | 7010 | 7460 | 7910 | | | | | | |

**Metric Size Radial Drive Wheel Tires**

| | | Tire Inflation (PSI) | | | | | | | | | | | | | | |
|---|---|---|---|---|---|---|---|---|---|---|---|---|---|---|---|---|
| Tire Size | Tire Config. | 6 | 7 | 9 | 10 | 12 | 13 | 15 | 17 | 20 | 23 | 26 | 29 | | | |
| 420/80R46 | Singles | NR | NR | 3520 | 3860 | 4080 | 4400 | 4680 | 5360 | 5840 | 6400 | 6800 | 7150 | | | |
| | Duals | 2560 | 2820 | 3100 | 3400 | 3590 | 3870 | 4120 | 4720 | 5140 | 5630 | 5980 | 6290 | | | |
| | Triples | 2390 | 2620 | 2890 | 3170 | 3350 | 3610 | 3840 | 4400 | 4790 | 5250 | 5580 | 5860 | | | |
| 710/70R38 | Singles | NR | NR | 2900 | 3150 | 3350 | 3650 | 3875 | 4375 | 4875 | 5300 | | | | | |
| | Duals | 2140 | 2330 | 2550 | 2770 | 2950 | 3210 | 3410 | 3850 | 4290 | 4665 | | | | | |
| | Triples | 1995 | 2175 | 2380 | 2585 | 2745 | 2995 | 3180 | 3590 | 4000 | 4345 | | | | | |

NR – Not Recommended

For Metric Equivalents:

- bar inflation pressure = psi x 0.069
- kg tire load limit = lb x 0.4535

*Table 1 — Load and Inflation Values for Radial Tractor Tires at Maximum Speed of 25 mph (40 km/h)*

## Monitoring Tractor Performance

Wheel slip, engine speed, and ground speed should all be closely monitored when field operations are performed that load the tractor close to a traction or power limit.

Wheel slip should normally be in the range of 8–12% to achieve peak tractive efficiency. It should not exceed 15–20% in peak overload situations. This provides good traction while acting as a cushion for the engine and drive train to soften impact of sudden overloads.

A radar monitor is recommended to precisely measure wheel slip. However, wheel slip can also be measured fairly accurately as follows: Mark a spot on the ground and a chalk mark on one rear tractor tire. Then drive the tractor under load with the implement in its normal operating mode for 10 complete rotations of the rear tire. Place another mark on the ground. Repeat the trip without the implement and again count wheel rotation between the two marks. Estimate the fraction of the last rotation as nearly as possible.

Check the number of rotations counted on the second trip, using Table 1 to determine the percentage of rear–wheel slippage. If less than 8-1/2 rotations are counted, add weight. If more than 9 rotations are counted, remove weight from the rear wheels.

If unable to measure tire slippage by counting wheel rotations, the tire–tread pattern produced when pulling under load provides an approximate indication.

When too much weight is used, the tire tracks will be sharp and distinct in the soil. There is no evidence of slippage. The tires are figuratively geared to the ground, and do not allow the flexibility of engine operation obtained when some slippage occurs.

| Rotations | Rear–Wheel Slippage Percent | What to Do |
|---|---|---|
| 10 | 0 | Remove Ballast |
| 9-1/2 | 5 | |
| 9 | 10 | Proper Ballast |
| 8-1/2 | 15 | |
| 8 | 20 | Add Ballast |
| 7-1/2 | 25 | |
| 7 | 30 | |

*Table 2 — Rear-Wheel Slippage*

Too much weight increases the engine power needed to move the tractor through the field and reduces the power available for pulling the implement.

When the tires have too little weight, they lose traction. The tread marks are entirely wiped out and forward progress is slowed.

When the tires have proper weight, a small amount of slippage occurs. The soil between the cleats in the tire pattern is shifted, but the tread pattern is visible (Fig. 2). Proper weighting allows the engine to perform at its best with maximum flexibility for varying loads.

If slippage exceeds 15 percent, take immediate action to reduce spinning. Slippage under 10 percent is barely visible. So, as a rule of thumb, if slippage can be seen, it's too much.

- Engine Speed. The engine should operate in the speed range specified by the manufacturer. Under normal conditions at full throttle, the speed should be near rated rpm but may drop a few hundred rpm for short durations when operated under high draft load. The operator may also shift up and throttle back if this does not cause the engine to labor. Follow the tractor manufacturer's recommendations.
- Ground speed of 5 mph (8 km/h) or higher is preferred, but no less than 4 mph (6.4 km/h) continuously.

A radar monitor is recommended for precise measurement of ground speed. Follow the tractor manufacturer's recommendations.

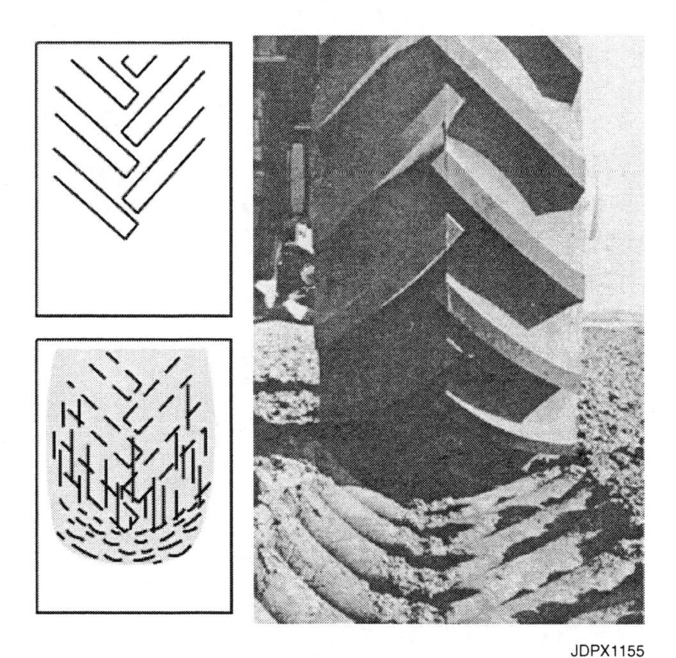


*Fig. 2 — Proper Weight on Tires*

If the tractor maintains engine and ground speed within the recommended limits but wheel slip is high, one or more of the following should be done.

- Reduce draft by reducing implement working depth or width.
- Add ballast but maintain correct tractor weight split.
- Consider larger diameter tires or dual wheels.

If the tractor is able to maintain a minimum of 4 mph (6.4 km/ h) and the wheel slip is within an acceptable range, reduce implement draft by reducing depth or width.

### Power Hop

Under high drawbar loads in certain soil conditions, MFWD and 4WD tractors may experience simultaneous loss of traction and bouncing, pitching ride. Occurrence of these two conditions is termed power hop or wheel hop. Power hop may become so severe that the operator loses ability to safely control the tractor. To regain control, engine speed and/or implement draft should be reduced until power hop subsides. However, these actions are only a temporary measure. To permanently control power hop, the tractor ballast and tire inflation pressures should be adjusted as previously described.

Power hop is strongly influenced by tractor weight split, amount and distribution of liquid ballast on each axle, and tire inflation pressures. If power hop is still present after following previously described procedures, front inflation pressures should be raised in 2 psi (0.14 bar) increments until the condition subsides or maximum tire pressure is reached.

**IMPORTANT: IMPORTANT: When raising tire inflation pressure, do not exceed the tire manufacturer's recommended maximum inflation. Make sure that all tires on an axle are inflated to the same pressure.**

If power hop cannot be eliminated with these procedures, consult the tire or tractor manufacturer for additional guidance.

## Implement Depth Control

Implements can be equipped with a wide range of depth control systems. These range from simple mechanical depth stops installed on the rod of a hydraulic cylinder (Fig. 3, left) to modern monitoring systems that allow the tillage depth to be displayed and adjusted in the tractor cab during operation.

Mechanical Depth Stop

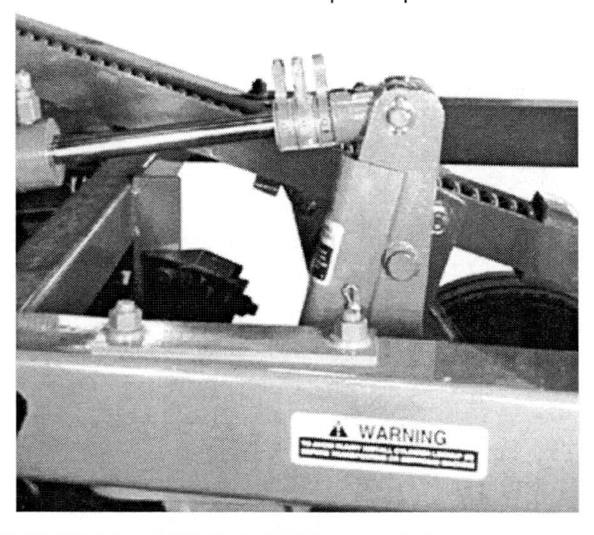


Crank–Controlled Depth Stop

JDPX6074, 6075

*Fig. 3 — Manual Implement Depth Stops*

On many implements, the depth control adjustment of the entire machine can be made from one single location. In the system on the right in Fig. 3, depth control is adjusted by lowering the machine to the desired operating depth and turning the crank until the stop bracket contacts the valve plunger. When the operator lowers the implement in the field, the machine will lower until the stop bracket contacts the valve plunger, shutting off the flow of oil to the hydraulic cylinders.

Some tractors allow for the implement depth to be set from inside the cab. In Fig. 4, the SCV controls located in the cab can be used to set up an upper and lower operating limit. A single height sensor mounted on the implement monitors implement height. Moving the SCV control lever forward or backward into the detent position will cause the implement to automatically raise or lower to the preset limit.

Many operating tillage implements use an even more sophisticated system for depth control. Using a combination of multiple height sensors, electronic control units, and hydraulic proportioning valves, precise depth control can be achieved during field operation.

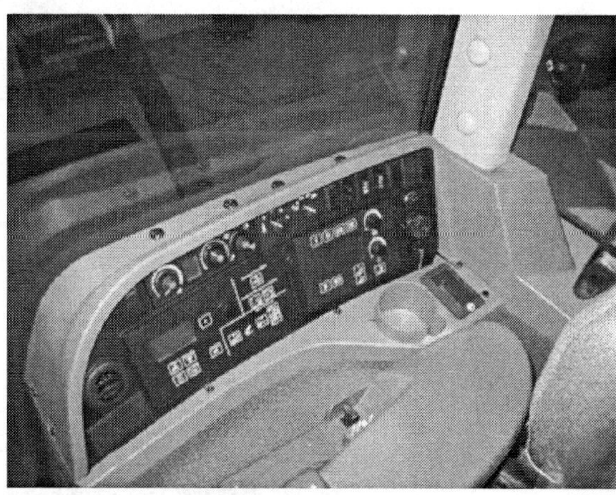

JDPX6076

*Fig. 4 — SCV Controls Upper and Lower Operating Limit*

The depth control system in Fig. 5 includes a separate display in the tractor cab to monitor depth position, as well as a separate electronic control unit (Fig. 6) mounted on the implement. Each section of the implement can be monitored and adjusted independently of the other sections. A hydraulic proportioning valve (Fig. 7) is mounted on each cylinder of the implement to regulate hydraulic oil flow. A height position sensor (Fig. 8) is also mounted at each wheel location. A five–section implement can have six hydraulic proportioning valves and six height sensors.

JDPX6077

Fig. 5 — Display Inside Cab Monitors Depth Position

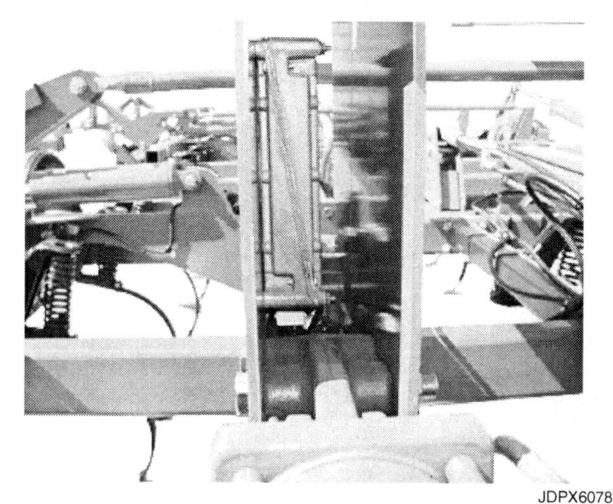

JDPX6078

Fig. 6 — Electronic Control Unit Mounted on Implement Monitors Depth Position

JDPX6079

Fig. 7 — Hydraulic Proportioning Valve Mounted on Each Cylinder of Implement

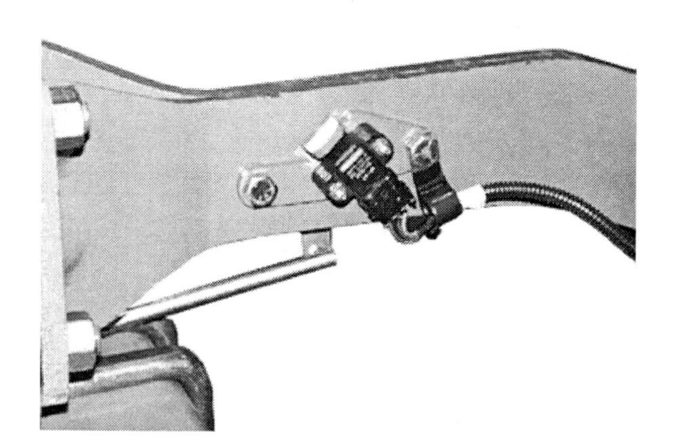

JDPX6080

Fig. 8 — Height Position Sensor Mounted on Each Wheel Location

As the implement moves through the field, the height sensors constantly monitor the height position and send this information to the control unit. If depth adjustment is required, the control unit activates the tractor SCV as well as the individual hydraulic valve at the cylinder until that section of the machine raises or lowers to the preset depth set by the operator. This process continues constantly for each section as the implement moves through the field to ensure precise depth control across the full width of the implement. The tillage depth for each section is shown on the display in the cab. The operator can make changes to the tillage depth from the cab by pressing a button on the display. If desired, the operator can adjust each section of the machine to a different depth to accommodate field positions.

## Tillage Implement Setup and Adjustments

A little extra care in adjusting a tillage implement for operation is time well spent. Properly adjusted implements not only do a better tillage job but save time and fuel in the field and reduce downtime for repairs and field adjustments.

Considering the large number of types and designs of tillage tools available, specific suggestions for setting up and adjusting each tool is far beyond the scope of this discussion. Rather, instructions in the operator's manual for each tool should be carefully followed to obtain optimum performance in the field.

Most problems with tillage tool performance are caused by improper adjustment or faulty components. When the tillage tool does not perform satisfactorily, always refer to the operator's manual to be sure the tool has been adjusted and components maintained according to instructions. If the trouble persists after following assembly, operating, and adjusting instructions, consult the local implement dealer or manufacturer for additional guidance.

Tillage tool performance problems can often be prevented with simple maintenance operations at the beginning of each season. These include:

1. Lubricate according to instructions. Clean grease fittings to avoid forcing dirt into bearings.
2. Clean, inspect, and lubricate or repack wheel and coulter bearings. Replace as needed.
3. Examine hydraulic hoses, couplings, and cylinders for wear, damage, or leaks. Repair or replace as needed.
4. Check for loose or missing bolts and nuts. Replace worn or broken parts.
5. Replace worn, dull, or cracked soil engaging components such as sweeps, shares, disks, or coulters.
6. Check and replace bent or cracked components such as standards or beams.
7. Check alignment of soil engaging components. Level the implement from side to side and fore and aft. Measure vertical distance from point to frame and distance between ground engaging tips. If any measurement is not essentially equal, consult the operator's manual for corrective action required.
8. Check operation of safety trips or reset mechanisms to be sure they function freely. Refer to the operator's manual for proper method of checking and adjusting.
9. Make certain all tires are inflated to the recommended pressure to provide level machine operation.

## Transport and Safety

Safety is a prime consideration in the design and manufacture of equipment, but there is no safety device that can replace a careful operator. Follow these rules and the specific steps listed in the operator's manual for safety and welfare of the operator and bystanders.

1. Match equipment to the tractor. Attempting a job that is too big for the tractor is self–defeating. Dangerous mismatching usually occurs when a tractor is undersized for the tillage tool. Fortunately, serious mismatching of tillage equipment is rare because it is usually obvious to the operator.
2. Provide proper tractor ballast and static weight split, as previously discussed, for tractor stability. Never pull from a point higher on the tractor than the recommended hitch point.
3. Match hydraulic connections. A serious error is to interchange hose ends on the auxiliary cylinder so the control valve operates in reverse. When the control is pulled for raising, the implement goes down, a real hazard. After hydraulic hoses are attached between tractor and implement, carefully test for proper coupling. See hydraulic warning below.
4. Check implement position before hitching it to the tractor. The implement should be parked on a firm, flat surface. Block mounted implements securely before lowering the hydraulic lift. Position the tractor directly in front of the implement before starting to back toward it for hitching (Fig. 10).

Hitching is much easier with an assistant. If no assistant is available, a snap–close hitch, self-centering hitch, or telescoping–tongue can ease hitching.

5. Use forethought in unhitching. Use hydraulic lift arms to support integral hitches. Don't try putting hitch pins in place while the tractor is in gear. Set brakes or put the transmission in park before getting off the tractor.

### Safety When Working With Hydraulic Fluid

JDPX5255

**CAUTION: Escaping fluid under pressure can penetrate the skin causing serious injury. Relieve pressure before disconnecting hydraulic or other lines. Tighten all connections before applying pressure. Keep hands and body away from pinholes and nozzles which eject fluids under high pressure. Use a piece of cardboard or paper to search for leaks. Do not use your hand.**

**If ANY fluid is injected into the skin, it must be surgically removed within a few hours by a doctor familiar with this type injury or gangrene may result.**

**CAUTION: Do not feel for pinhole leaks. Escaping fluid under pressure can penetrate skin causing injury. Relieve all hydraulic pressure before working on a pressurized hydraulic line or component.**

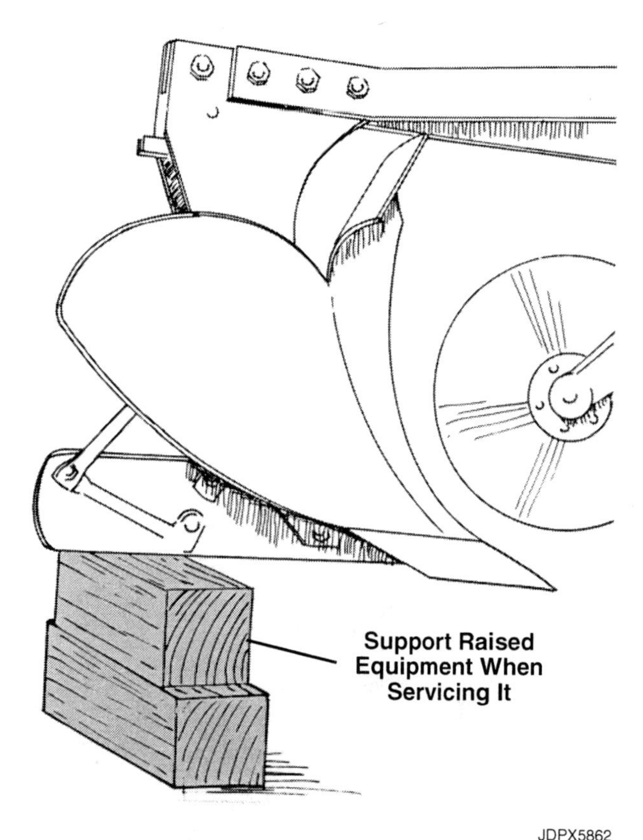

*Fig. 9 — If a Machine Must Be Serviced in the Raised Position, Use Transport Locks and Block It Up*

6. Don't permit people between tractor and tillage tool, especially when backing up to hitch. A foot may slip off the clutch or the implement may move, catching someone between the tractor and implement.
7. Always lower the implement to the ground or use jacks, blocks, transport links, or lock pins when it's not in use or when working on the machine. If the equipment must be serviced in the raised position, use blocks, jack stands, or other support on firm ground (Fig. 9). Never depend on the hydraulic system to hold the tillage tool up.
8. Pins used to connect the tillage tool to the tractor should be the proper size and secured with a clip or pin. Don't use makeshift pins such as long bolts, which can break, bend, or jump out under load. Use safety hitch pins that are easy to remove but have springs or clips to keep them in place (Fig. 10). Three–point hitch links should also be secured with restraining clips or pins.

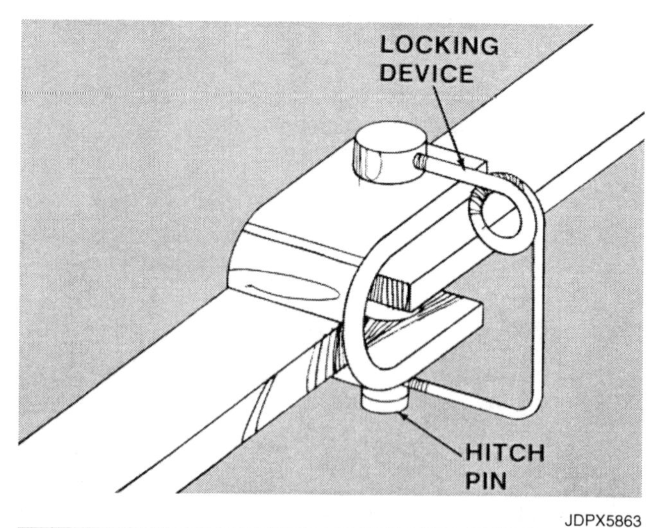

*Fig. 10 — Use a Locking Hitch Pin to Prevent Accidents*

9. Never carry riders on the tractor or permit others to ride on the tillage tool.
10. Reduce speed when transporting over rough or uneven terrain.
11. Check wings or outrigger locking mechanisms on fold-up tillage equipment. They can fail and let the equipment fall during transport.
12. Use proper lights, reflectors, and a clean slow–moving vehicle (SMV) emblem when transporting equipment on road or highway.
13. When transporting, put the machine in as narrow a configuration as possible. Most wide equipment has a special transport position. Some states require a special permit to transport equipment wider than a specified width. Check local and state regulations.

## Summary

Traction, flotation, and soil compaction are closely related. Changes in soil–surface conditions, soil contact area, and each vehicle weight directly affect all three.

Major items for setting up a tractor for tillage are (a) total tractor weight and weight distribution on front and rear axles, (b) type of ballast used (cast and/or liquid weight), and (c) tire inflation pressures. A properly set–up tractor will provide improved traction and flotation, reduced compaction, improved ride, reduced tire wear, improved stability, and better control of power hop.

Wheel slip, engine speed, and ground speed should be closely monitored under high–load field operations. Wheel slip should be in the range of 8–12%, engine speed should be within the range specified by the manufacturer, and ground speed should be maintained at over 4 mph (6.4 km/h), preferably at 5 mph (8 km/h).

Most problems with tillage tool performance are caused by improper adjustment or faulty components. Operator manuals provide specific suggestions for proper adjustment and safe operation for individual tillage tools.

## Test Yourself

### Questions

1. What three factors are included in proper setup of tractors for tillage?
2. What is traction as applied to agricultural equipment? What is flotation?
3. List at least four benefits from a properly set-up tractor.
4. What is the recommended static weight split for front and rear axles of 4WD tractors used for towed implements?
5. (T/F) Pulling a lighter load at higher speed can reduce wheel slip without increasing soil compaction.
6. What three factors must be monitored when the tractor is operated at loads close to a traction or power limit?
7. (Select one.) Wheel slip for optimum tractive efficiency should be (a) 8–12%, (b) 15–20%, (c) 1–5%.
8. (Fill in blanks.) Most problems with tillage tool performance are caused by ________________ or ________________.
9. Why should care be taken not to interchange hydraulic hose ends when connecting a tractor with a tillage implement?
10. (Select one.) Alignment of soil engaging components should be checked (a) only when the implement is first assembled, (b) at least every six months, (c) before each season.
11. (T/F) A large, heavy-duty bolt can often be a satisfactory hitch pin.
12. (Fill in blank.) The best safety device for tillage equipment is a careful ______________________.

# Toolbars

# 5

JDPX6082

*Fig. 1 — Toolbar Equipment Matches Individual Needs*

## Introduction

Toolbar tillage and planting began in the Southwest and later moved into the Mississippi Delta area. Only in recent years has there been much interest in toolbar equipment in the Corn Belt. With relatively small tractors and stable 40–inch row spacing (102 cm), there formerly was little incentive for Corn Belt farmers to use toolbars.

But as row spacings began to change, and planting and cultivation became 6–, 8–, 12–, and even wider operations, toolbars have taken on new importance in assembling the particular equipment each farmer needs (Fig. 1 and Fig. 2). Since the late 1960s the variety, size, and use areas of toolbar equipment have grown rapidly.

Because of the vast array of possible toolbar–equipment combinations, only some of the basic components and a few applications will be considered. Details will be minimal, and we will not pursue individual principles of operation, preparation, or trouble shooting, but will discuss some general areas of application.

JDPX6083

*Fig. 2 — Unit Planters Are Common Toolbar Attachments*

**IMPORTANT: Every tractor has certain lift and draft limitations, and all ready–made equipment is likewise designed and constructed to meet definite strength and performance specifications. This permits manufacturers to provide recommendations for matching of implements to specific tractor models.**

However, most toolbar–based implements assembled by dealers or farmers have not undergone the extensive testing of factory–built machines. Therefore, it may be difficult to predict the exact size of bar required, the correct number of clamps, or the lift capacity required to transport the assembled machine. Too much weight can overload the tractor hydraulic system, reduce tire life, and cause poor tractor stability. Excessive draft means unsatisfactory field performance and possible tractor or equipment damage, and could cause tractor instability.

Too much power for the strength of the implement could result in serious equipment damage, especially in adverse operating conditions.

The best method for determining toolbar size and tractor capacity is to compare plans for combinations with similar toolbar implements offered by the manufacturer for certain tractor models. It also is helpful to check with farmers who may be successfully using an assembly similar to what is being planned.

## Toolbars

For many years toolbars were solid–steel bars, 2, 2-1/4, or 2-1/2 inches (50, 57, or 64 mm) square. Later, square hollow bars became available to reduce weight by 55 to 60 percent, cut cost by as much as 20%, and retain ample strength for many small jobs. These also are 2 to 2-1/2 inches (50 to 64 mm) square and often are used interchangeably or in combination with solid bars.

A hollow 3-1/2 x 7 inch (89 x 178 mm) flattened–diamond shape is used for some toolbars. This shape provides considerably more strength than small, square toolbars, and permits use of some of the same attachments by using longer bolts.

Many rear–mounted row–crop cultivators use 4 x 4 or 5 x 7 inch (100 x 100 or 127 x 178 mm) hollow bars for maximum strength per pound. However, these bars are seldom used for other applications, so they will not be discussed further here.

Box–beam toolbars provide outstanding strength–to–weight ratio for large tillage and planting units. These bars range from 4 x 7 to 7 x 7-1/2 inches (100 x 178 to 178 x 190 mm) for different applications.

The type of toolbar must be decided on the basis of anticipated use and total weight. Oversized toolbars provide added strength, but increase weight and cost. However, replacement of toolbars that fail due to insufficient strength may increase total costs even more.

## Toolbar Hitches

One–piece integral toolbar hitches may be used with solid or hollow bars of various lengths, depending on the expected weight and draft load. A second bar may be added, by using spacers, for extra strength or clearance. The hitch may be removed from the bar and reattached to a different bar with other attachments, if desired, to eliminate the need for completely changing all equipment.

A three–piece hitch attached directly to the toolbar provides conformity with different hitch dimensions. Relocating brackets and switching pins permits operation with tractor hitches of various sizes.

Increased strength is provided by doubling bars and using straddle–mounted hitch pins. Additional rigidity is provided by mast braces, truss rods, and diagonal braces between bars. Bars can be spaced at different distances for easy placement of a variety of attachments.

Box–beam toolbars require extremely rugged hitches to match the strength of large tractors and the weight of large implements. Some hitches are attached directly to the box–beam bar. Additional versatility is provided for special applications by mounting hitch brackets on a separate bar clamped in front of the box–beam toolbar (Fig. 3). Mastside plates can be moved laterally for maximum flexibility in clamping attachments to the main bar. Shifting the mast and hitch clamps back on the toolbar and using spacer clamps (Fig. 4) permit attachment of an optional square bar.

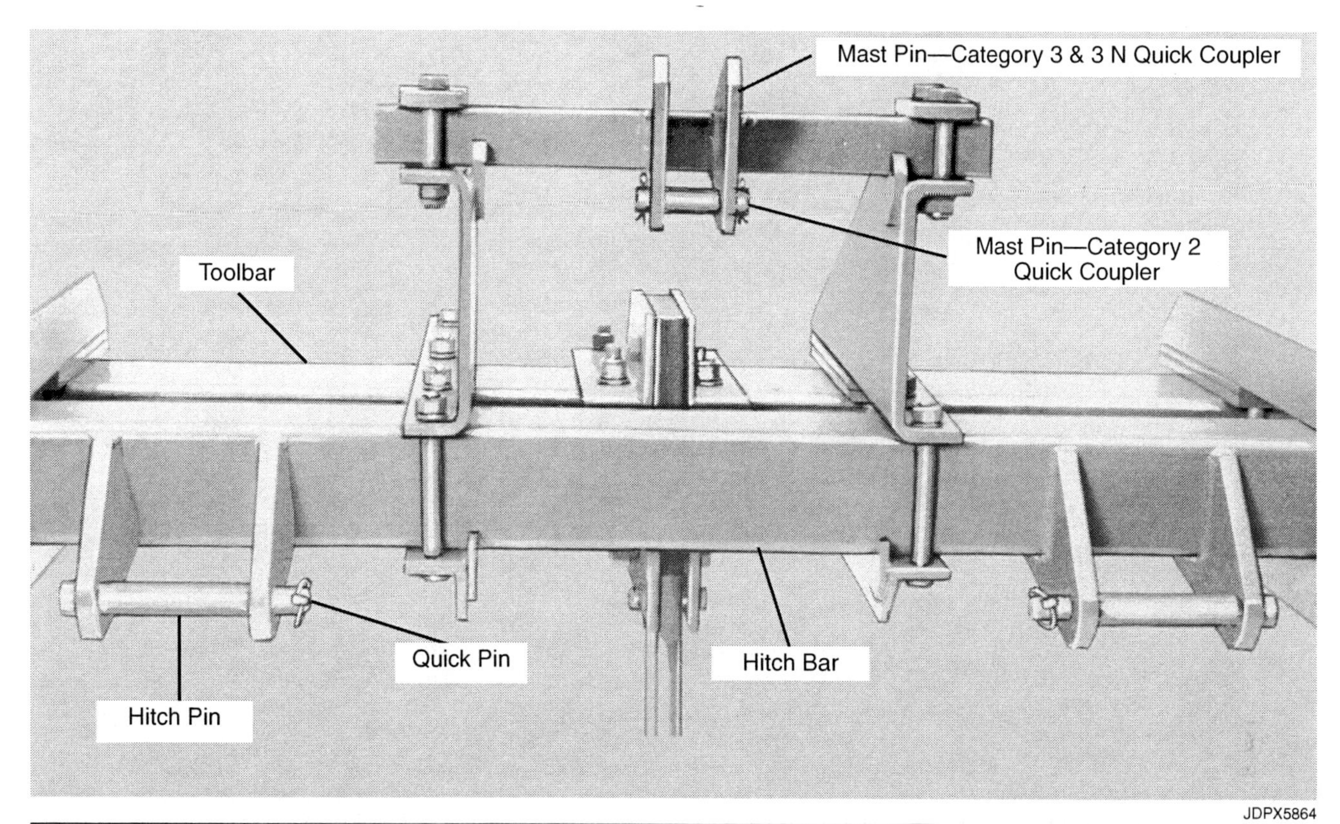


*Fig. 3 — Versatility and Adjustability Are Keys to Toolbar Success*

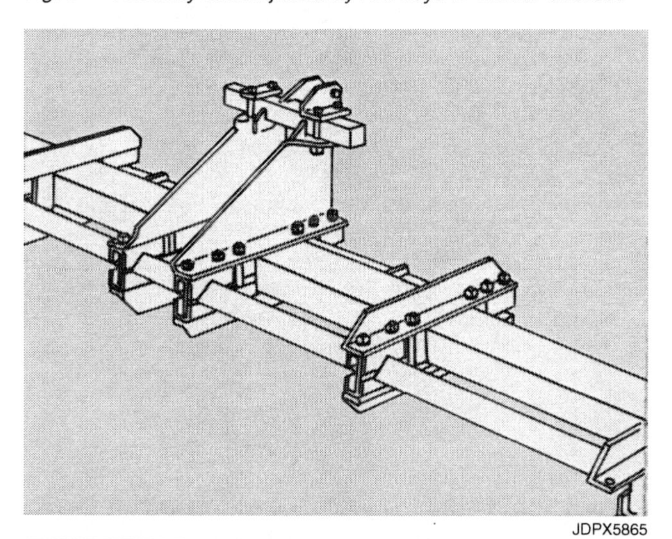


*Fig. 4 — Relocating Carrier on Toolbar Permits Simple Addition of Second Bar*

If heavy attachments (such as subsoilers) are mounted on the box–beam bar, bedder bottoms or other attachments may be clamped to a second bar attached to the box–beam by flexible linkage. This permits regulation of working depth of the front bar by the tractor load–and–depth system or gauge wheels, while the rear bar is controlled by its gauge wheels, providing independent depth control for front and rear units.

## Sled Carriers

Sled carriers can be the basic implements for a complete specialized–crop–growing system. With bed–shaping attachments (Fig. 5), they form precision beds in many different row spacings and configurations. Shapers are available for wide and narrow beds, various widths and depths of furrow, and salt–ridge shapers for irrigated areas where salt may damage crops (Fig. 6).

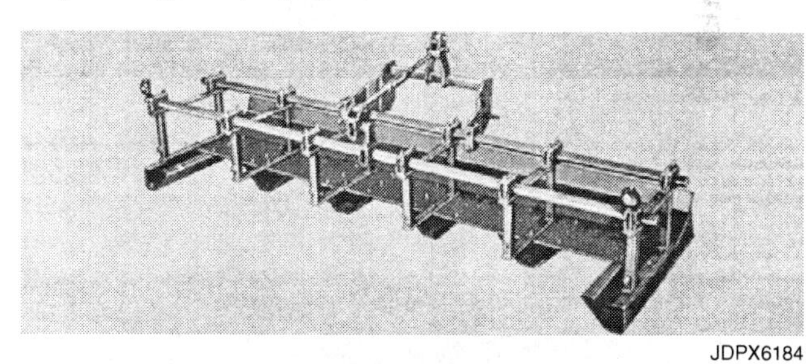


*Fig. 5 — Sled Carriers Can Make Precision Beds*

Attaching unit planters (for beets, beans, vegetables, or cotton) permits shaping beds and planting in one simple operation (Fig. 7). Bed shapers sweep dry topsoil from the beds, and seeds are placed directly in moist soil for rapid germination. Shapers can also plane off weeds that may have started since the field was bedded, and shape the beds for efficient application of herbicides.

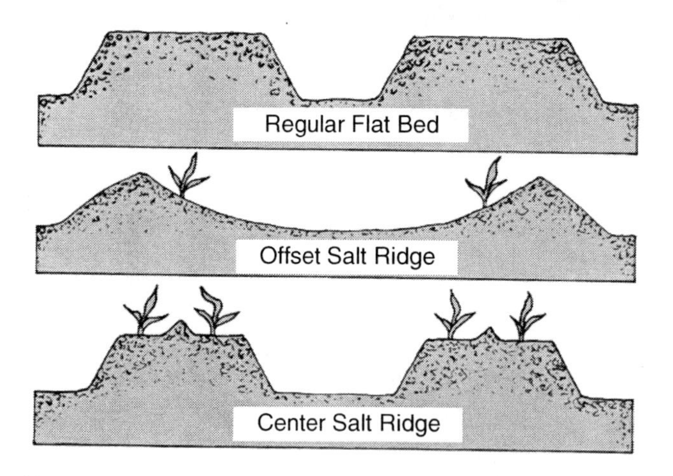

*Fig. 6 — Many Different Bed–Shaping Attachments Are Used*

*Fig. 7 — Once–Over Bed–Shaping and Planting Saves Time*

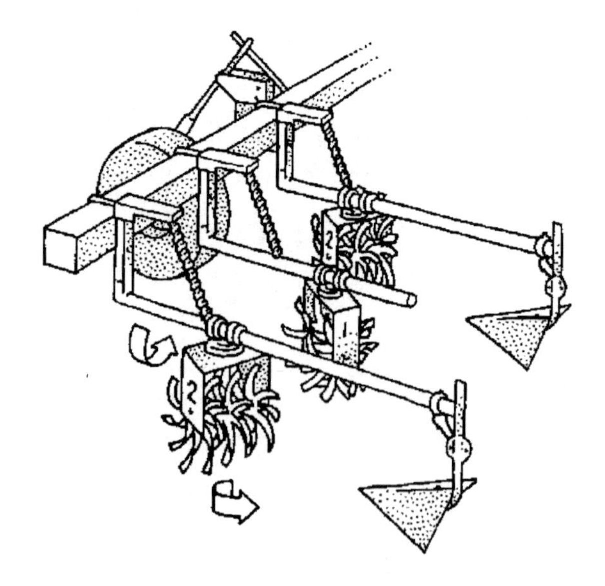

*Fig. 8 — Sled Carrier Can Provide Precision Cultivator Guidance*

Furrows between beds are uniformly shaped and compacted for optimum flow of irrigation water. They may also be used to guide cultivator attachments for later weed control (Fig. 8). Replacing unit planters and bed shapers with cultivating tools on the same sled carrier permits fast, precision cultivation with minimum plant damage and operator fatigue.

### Spacer Clamps

Rigid spacer clamps permit attachment of a square second bar to box–beam toolbars, or for clamping two, three, or four square bars together. Many clamp configurations are available to space two bars from 6-1/2 to 20 inches (165 to 508 mm) apart. Additional clamps of equal or different size may be used to attach a third or even fourth bar for individualized assembly.

### Gauge Wheels

Toolbar gauge wheels are used to control implement working depth or toolbar height for specific operations. They are generally available in single or dual wheel configurations.

A hydraulically controlled lift–assist wheel is also available for extra–heavy loads, such as fertilizer hoppers.

### Standards and Shanks

The variety of toolbar shanks and standards is almost endless. They range from subsoiler standards (Fig. 9) that can work 24 inches (610 mm) deep, to vegetable–cultivating standards for weed knives, small shovels, etc. (Fig. 10). In between are shanks for chisel points or sweeps (Fig. 11) and standards for lister or bedder bottoms. Combination units feature chisel points and bedder bottoms. There are also irrigation border disks, disk–bedder standards with or without subsoilers, and flexible–frame adjustable 2– or 3–blade disk gangs, plus many other options.

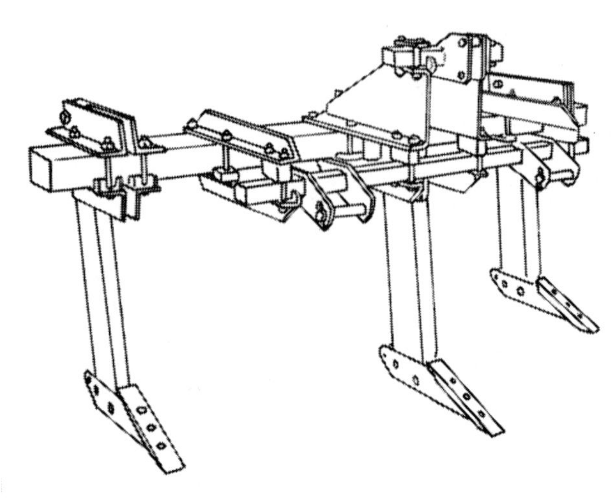

*Fig. 9 — Subsoiler Standards Usually Clamp to Box–Beam Bars*

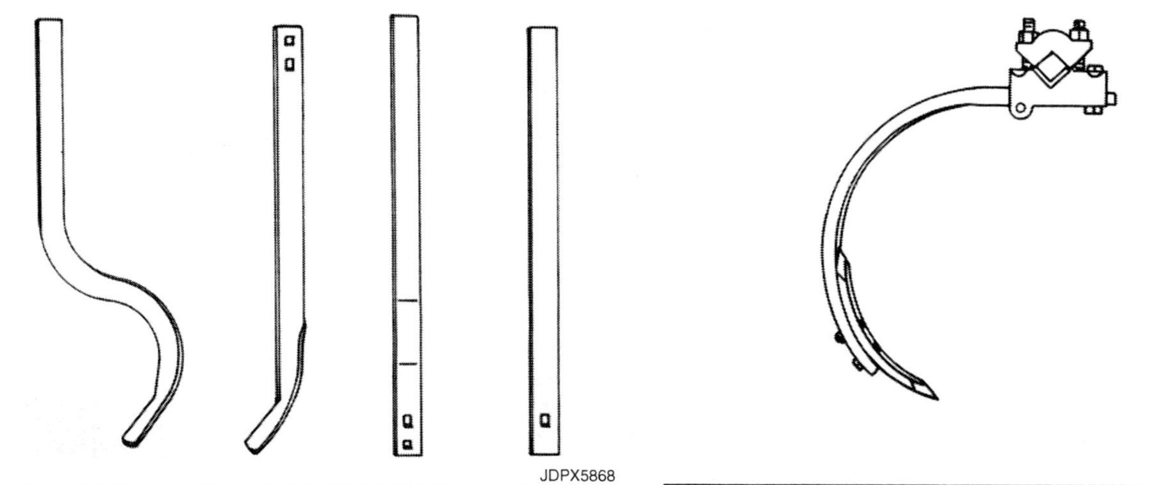


*Fig. 10 — Vegetable–Cultivating Standards Fit Different Tools and Row Arrangements*

*Fig. 11 — Chisel Plow–Type Shanks Fit Square Bars*

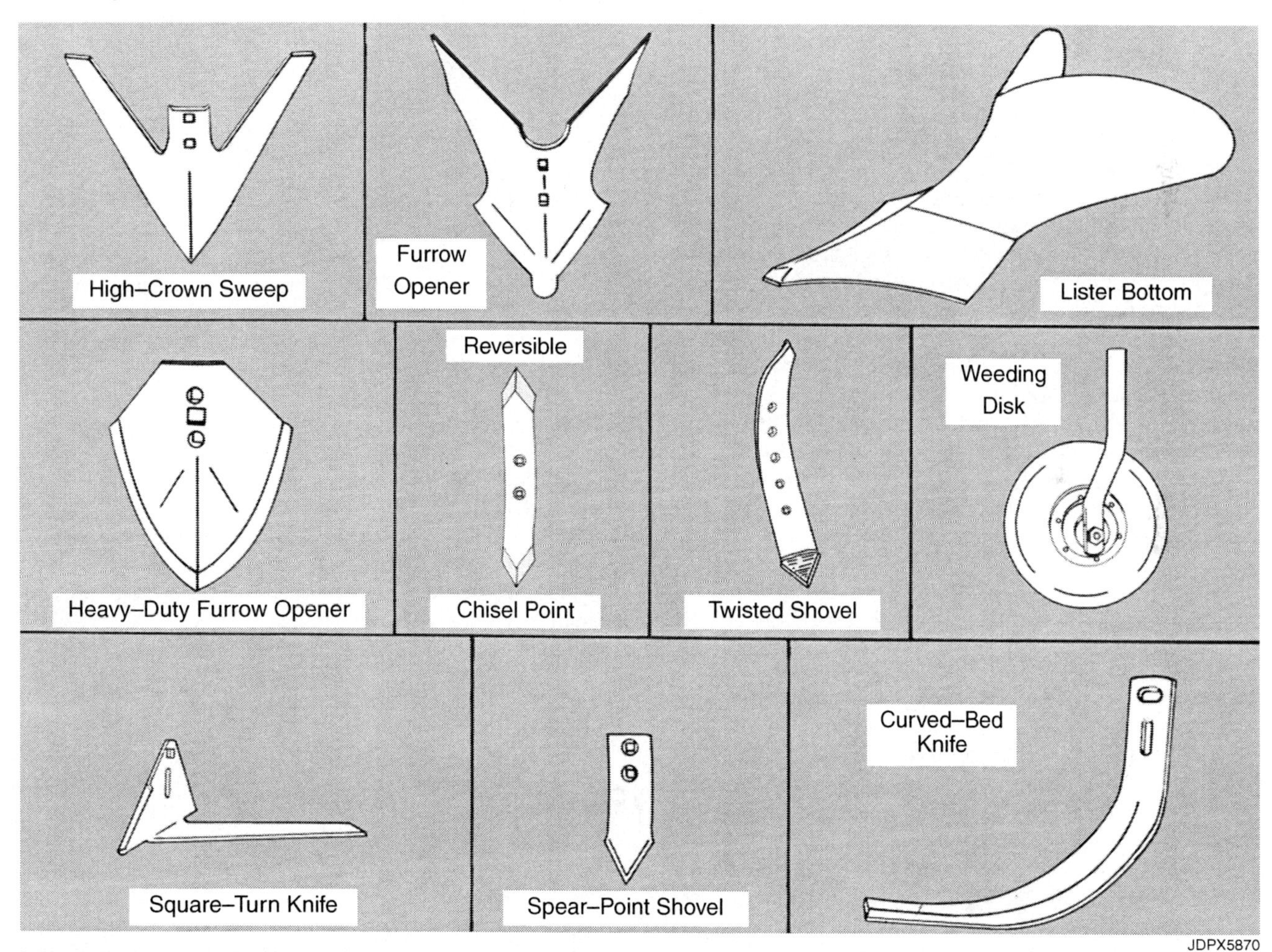


*Fig. 12 — Almost Any Type of Soil–Engaging Tool Can Be Used With Toolbars*

## Soil–Engaging Tools

Toolbars may be equipped with soil–engaging tools to match almost any soil, crop, or tillage condition (Fig. 12). They can use the full line of chisel–plow sweeps, shovels, furrow openers, reversible spike and chisel points, and twisted shovels. Also available, with different standards, are various sweeps, points, shovels, disk hillers, furrow openers, weed knives, etc., used for row–crop cultivation.

Hundreds of combinations are possible when sweeps, shovels, and disks are used together for special bedding or cultivating operations, provided tractor power, lift capacity, and stability are not exceeded.

### Markers

Markers are used to guide the operator where there is a possibility of overlapping or skipping. Marker length is usually adjustable for a range of row widths or equipment sizes.

Markers may be raised automatically, by chain and sheave attachment on the tractor drawbar, or hydraulically. Selective lowering may be accomplished automatically as the toolbar is lowered or by hand–releasing the proper latch. An optional electric release for tractors with cabs is available for some markers.

### Toolbar Transport

Most toolbars up to about 15 or 16 feet (4.5 or 5 m) are transported on the tractor 3–point hitch without change. If length is much greater, some means of reducing width is usually required to permit safe transport. An endways transport attachment reduces width and permits easy movement with maximum tractor stability. Toolbar gauge wheels may sometimes be removed and used as transport wheels. The towing hitch for endways transport is equipped with a cross shaft to fit the tractor lower links or quick coupler for easy hookup and raising for transport. Hydraulic cylinders mounted inside the toolbar control folding of some large toolbars (Fig. 13). For transport, a typical 32–foot (9.7 m) bar may be reduced to 21 feet (6.4 m), which still requires considerable care to avoid collisions or running tractor wheels too close to the road edge. Only a small area near each hinge restricts placement of attachments.

### Summary

To catalog all possible and practical combinations of soil–conditioning, cultivating, and planting combinations for hitch-carried toolbars would be a difficult task. In fact, just listing toolbar attachments from various manufacturers is almost impossible. They include not only many types of soil–engaging tools but a wide variety of solid and box–beam bars, standards of various sizes and lengths, and such auxiliary equipment as various spacers, gauge wheels, lift–assist wheels, and markers.

JDPX6187

*Fig. 13 — Hydraulically Folded Toolbar*

Toolbars save money by providing a single basic frame for a wide choice of equipment. They are an ideal basic unit for innovative farmers who want to try combinations not available in complete implements sold by dealers. Some farmers have two or more toolbars, permitting them to change operations without tearing down one hookup to use another.

But caution is necessary. Implement manufacturers carefully check strength, and keep size and weight within the power, hydraulic lift, and stability capacity of recommended tractors. They cannot do this for farm–assembled toolbar combinations. It is up to users who experiment with toolbars to assemble combinations with sufficient overall strength, and to avoid overtaxing tractor power and stability and lift limits of 3–point hitches.

## Test Yourself

### Questions

1. What is one of the major problems encountered in do–it–yourself implement design with toolbar components?
2. How can toolbar implement weight be reduced for some lighter small–bar applications?
3. Name two reasons for using multiple toolbars on one tool carrier.
4. List five general categories of commonly used toolbar attachments, one specific operation for each category, and a crop on which the attachment could be used.

# Primary Tillage

6

## Introduction

As described in Chapter 1, tillage can be classified as primary or secondary, depending on depth of tillage and roughness in which the soil surface is left. Primary tillage is more aggressive, usually works the soil 6 inches (15 cm) deep or more, and leaves the surface rough.

Primary tillage can be further classified as inversion or non-inversion tillage. Inversion primary tillage completely inverts the soil to bury residue. It leaves the soil rough but with little or no surface residue. In contrast, non-inversion primary tillage works the soil to the same depth, but is intended to disturb the surface as little as possible. It leaves the surface rough and covered with most or all of the pre-tillage residue.

This chapter will include a discussion of several tillage tools commonly used for inversion or non-inversion tillage. The chapter is divided into two sections, one for each type of tillage.

## Inversion Primary Tillage

### Moldboard Plow

For thousands of years, plows have been used as primary tillage tools to prepare seedbeds and rootbeds for crops. Modern moldboard plows, as well as the crude plows still used in some underdeveloped parts of the world, have evolved from forked sticks pulled by primitive farmers or their animals.

Such Colonial leaders as Thomas Jefferson and Daniel Webster helped adapt European plows to American conditions. Those early cast-iron plow bottoms worked well in sandy and gravelly soil, but they wouldn't scour properly, if at all, in sticky Midwest prairie soil.

That problem prompted development of steel bottoms, which became an important factor in the settlement and agricultural development of the Prairie states. Prior to that time, some people had predicted that the prairie soil would have to be abandoned for growing crops because of the difficulty in plowing.

Frontier blacksmiths and farmers continually modified plow design for localized soil and crop conditions. Modern farm equipment manufacturers are continuing this evolution as plowing speed increases and designers work to reduce draft and improve plow performance.

JDPX6171

*Fig. 1 — Plowing Buries Trash and Aerates the Soil*

## What the Plow Does

The moldboard plow cuts, lifts, and turns the furrow slice (Fig. 1) and in so doing it:

- Buries some or all of trash and crop residue
- Aerates the soil
- Controls weeds, insects, and crop diseases
- Incorporates fertilizer into the soil
- Provides good seedbeds for better germination

Incorporation of plant residue (Fig. 2) and aeration of the soil by tillage also stimulate growth of microorganisms, which decompose trash and other organic materials. Accelerating decomposition of organic matter increases the supply of available nitrogen, phosphorus, potassium, and other plant nutrients. Soil microorganisms require the same favorable conditions of warmth, moisture, and aeration as are needed for quick seed germination.

Good plowing sandwiches organic matter between furrow slices to form a wick for better water absorption and storage and for faster decomposition of residue. It also increases soil porosity and provides more air for faster, stronger root growth.

Good soil pulverization in plowing reduces the cost of later tillage, but unless planting will start immediately, each furrow should retain a distinct crown. This helps reduce runoff, reduces ponding and puddling in low spots, reduces wind erosion, increases moisture infiltration, and speeds surface drying.

Plowing to the same depth each year may lead to a plow sole or hardpan, just below plowing depth, which can severely restrict root growth and water movement. Occasional chisel plowing or plowing several inches deeper every few years will help break the compacted layer.

JDPX6136

*Fig. 2 — Plows Incorporate Heavy Residue*

## Moldboard Plow Bottoms

The business end of the moldboard plow is the bottom, normally from 12 to 20 inches (305 to 508 mm) wide. It is essentially a three-sided wedge, with a landside and share cutting edge as flat sides, and the moldboard as the curved side. Each bottom is attached to a standard, which in turn is fastened to the plow frame.

The main plow-bottom parts are all attached to a common part called the frog (Fig. 3 and Fig. 4) and include:

- Moldboard
- Share
- Shin
- Landside

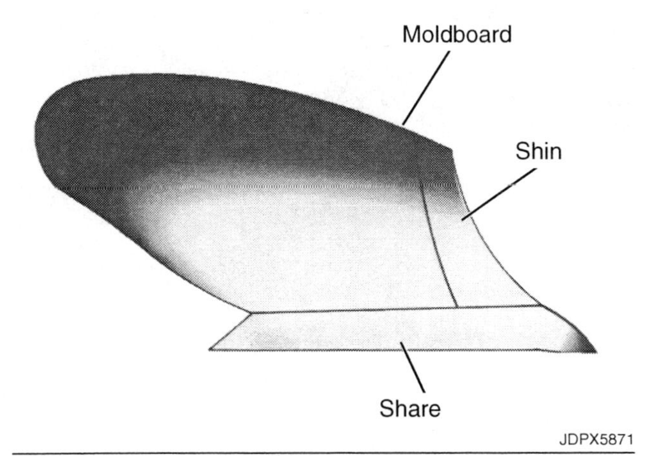

*Fig. 3 — Side View of Plow Bottom*

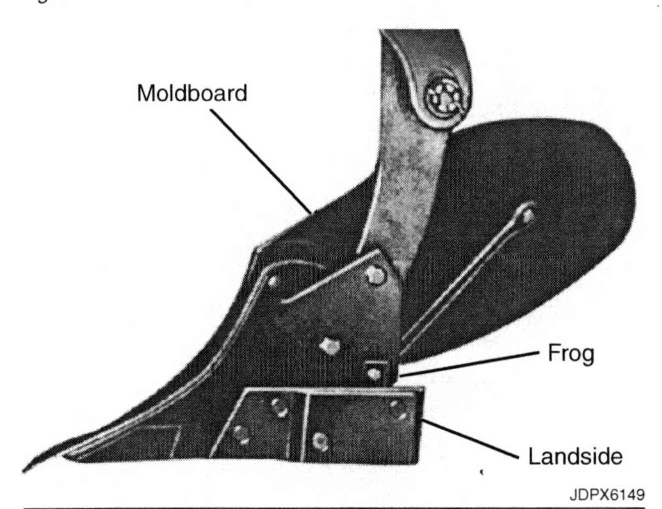

*Fig. 4 — Moldboard and Landside Attached to Frog*

## Soil Action on the Plow Bottom

The wedging action of the plow bottom moving through the soil exerts pressure upward and toward the open furrow. This turning causes blocks of soil to be sheared at regular intervals. This is the fracturing effect of a plow bottom under most soil conditions.

The blocks of soil rub and slip against each other as they move upward over the moldboard, causing granulation or crumbling of the furrow slice. As plowing speed increases, soil pulverization also increases, depending to a great extent on moldboard shape. Most of the granulation is done by the lower portion of the moldboard; the upper portion primarily turns the furrow slice.

## Levelness of the Plow Bottom

Plow bottoms are designed to run level and exert uniform pressure on the furrow slice. If a plow bottom does not run level, too much or too little pressure is exerted on the turning soil, resulting in poor plowing and poor granulation.

## Angle of the Furrow Slice

The angle or slope of the furrow slice is influenced by the speed of plowing, moldboard curvature, depth of plowing, and levelness of the plow.

## Plow-Bottom Design

Soil types and characteristics vary from light, abrasive sand to heavy, sticky clay. Depending on soil type, as well as climate and crops, soil moisture content when plowing may range from that of saturated rice paddies to extremely hard and dry soil, such as sun-baked sandy clay in irrigated desert areas. Soil physical condition may range from a loose, well-granulated structure to a hard, compact mass. This is affected by mineral, organic, and moisture content.

No single plow-bottom design can possibly do a satisfactory job in all soil conditions. For example, early plow bottoms were designed to be pulled slowly by animals, and didn't work well when pulled by tractors at higher speeds. Similarly, some current high-speed bottoms function poorly when operated at slower speeds. Draft varies widely between designs, and even between similar bottoms made by different manufacturers.

Hundreds of plow-bottom shapes have been made, each meant for a particular job. However, most present bottoms fail into six main types:

- General purpose
- High-speed
- Slatted
- Stubble
- Reversible
- Deep-tillage

Here's more about each type.

General-purpose bottoms (Fig. 5) have fairly slow-turning moldboards, which work well in sod and for faster plowing of old land, stubble, ordinary trash, and stalk cover. These bottoms are usually designed for speeds of 3–4 mph (4–6.5 km/h).

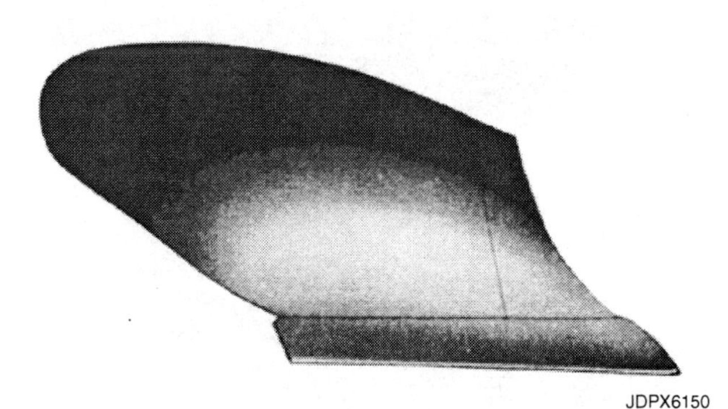

*Fig. 5 — General-Purpose Bottom*

High-speed bottoms (Fig. 6) have proved to be practical for plowing at 4–7 mph (6.5–11 km/h). The moldboard has less curvature at the upper end than general-purpose bottoms. Reasonably priced throwaway shares can be discarded when worn out.

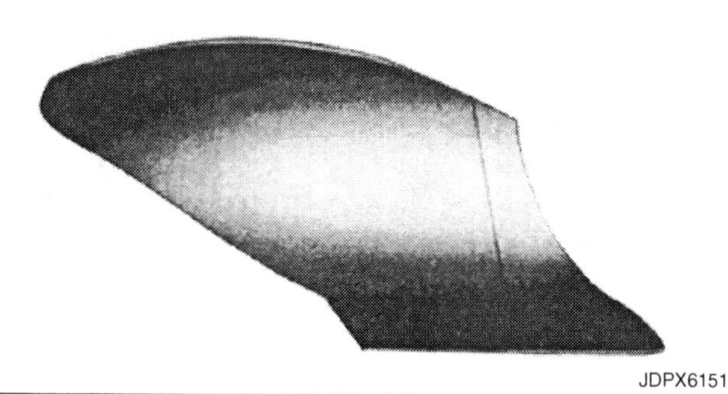

JDPX6151

*Fig. 6 — High-Speed Bottom*

Slatted moldboards (Fig. 7) scour better in difficult soil because about 50 percent of the moldboard has been removed. This concentrates the full pressure of the turning furrow slice on the remaining area, thus greatly improving scouring in waxy clay soils or loose, sticky soils.

JDPX6152

*Fig. 7 — Slat Bottom*

Stubble bottoms (Fig. 8) have an abruptly curved moldboard, which turns the furrow slice quickly. These bottoms are used primarily for difficult scouring conditions such as stubble land. The sharply turned moldboard thoroughly pulverizes soil, but doesn't lend itself to high-speed plowing. The limit is generally 2–3 mph (4–5 km/h).

JDPX6153

*Fig. 8 — Stubble Bottom*

Reversible bottoms (Fig. 9) have a unique design that provides even cutting in both directions. For two-way plowing, the moldboards are reversed hydraulically. They are used in any soil in which scouring is not a problem.

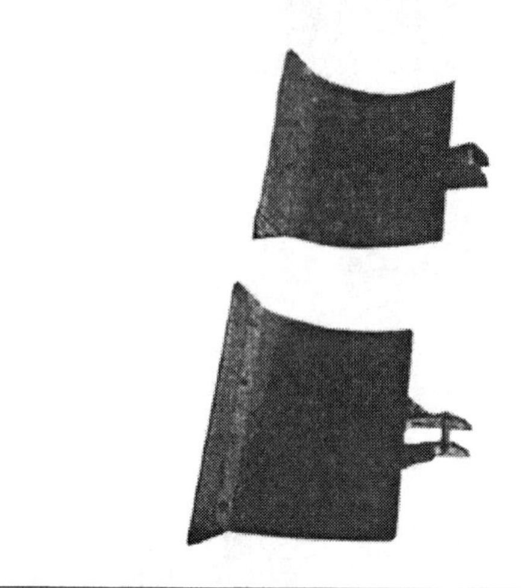

JDPX6154

*Fig. 9 — Reversible Bottoms*

Deep-tillage and semi-deep-tillage bottoms (Fig. 10) have high moldboards to permit plowing as deep as 16 inches (406 mm) in heavy soil. These bottoms are used most commonly in irrigated areas and in the South.

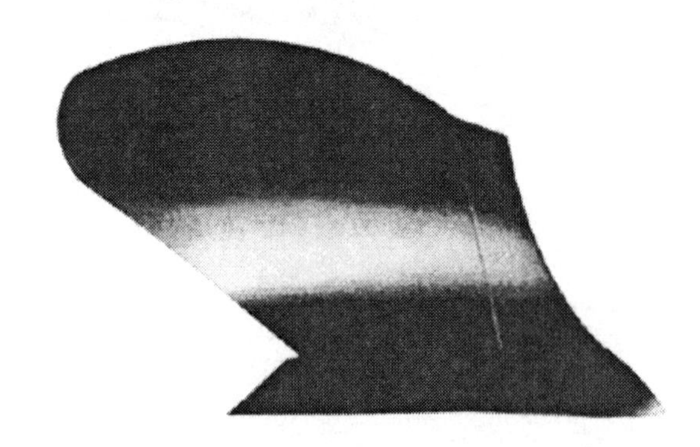

JDPX6155

*Fig. 10 — Semi-Deep-Tillage Bottom*

## Moldboards

The curved moldboard and shin of the plow bottom receive soil from the share and lift, fracture, and turn the furrow slice. The sliding action of the soil generates a great deal of heat and wear on the moldboard, and particularly on the shin, which therefore is replaceable on many bottoms.

## Plow Shares

The share is the business end of the plow bottom. It gives suction and penetration, and it cuts the furrow slice loose. Some lifting and a slight turning action starts at the share, but little granulation takes place. For many years plow bottoms had forged shares that could be resharpened when cutting edges and points wore dull. Now, however, farmers use throwaway or expendable shares that are simply discarded when worn. Several kinds of throwaway shares are manufactured to meet different plowing conditions.

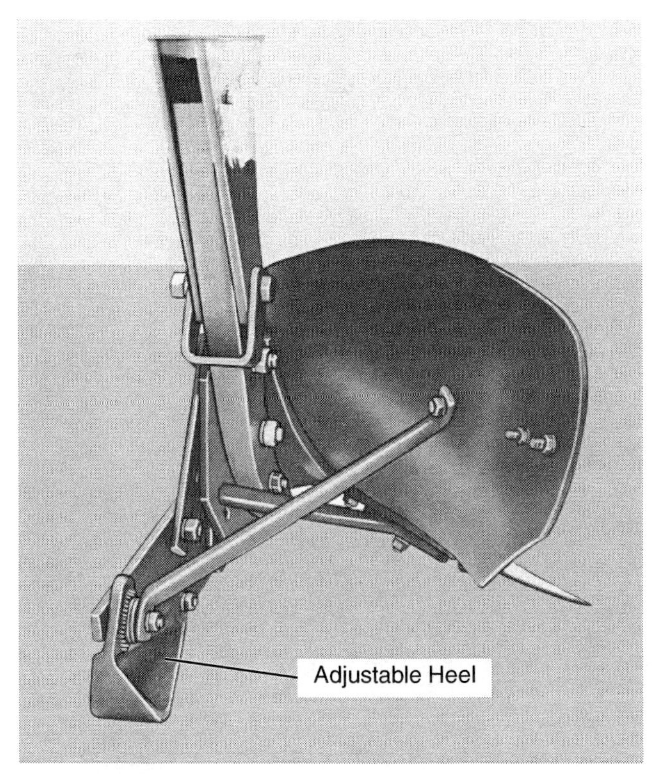


*Fig. 11 — Long Landside With Adjustable Heel*

## Moldboard Plow Landsides

The landside is a flat metal piece bolted to the frog, and it forms one side of the plow-bottom wedge. It helps absorb side forces from the turning furrow slice, steadies the plow, and helps keep the plow straight behind the tractor. Varying plowing conditions and plow designs require different landsides.

Steel landsides with reversible wear plates are used on high-speed bottoms in normal plowing conditions. The plate can be reversed for longer wear.

The long landside with adjustable heel is usually used on the rear bottom of integral moldboard plows (Fig. 11). The heel is adjusted vertically to mark the furrow bottom slightly and assist in controlling the rear of the plow.

A rolling landside is used on the rear bottom of some integral plows (Fig. 12). Rear wheels of some drawn and semi-integral plows are used as rolling landsides. Rolling landsides on some plows may be adjusted to increase or decrease pressure on the furrow wall to help make the plow trail straight behind the tractor.

*Fig. 12 — Rolling Landside in Operation*

## Moldboard Plow Standards

Each bottom is bolted to a standard assembly, which in turn is attached to the plow frame. Some older plows (Fig. 13) had one-piece beams and standards, formed from a web or I-beam member, and bent and flattened to the desired shape. Others were rigidly assembled from rectangular bar stock and flat steel plates. No release mechanism was provided on these standards or beams to protect bottoms from obstructions in the soil, though a safety-release hitch was used on some drawn plows of this type to disconnect the plow from the tractor if a bottom hit an obstruction.

Release-type standards hold the bottoms in position on most current plows, and provide protection for bottoms and frame from damage by rocks and stumps in the soil. Design of standards varies greatly, but four basic types are commonly used.

- Shear-bolt
- Safety-trip
- Hydraulic automatic-reset
- Spring automatic-reset

JDPX6156

*Fig. 13 — One-Piece Beam and Standard Used on Old-Style Plows*

Shear-bolt standards (Fig. 14) are economical protection where fields have very few buried obstructions.

JDPX6188

*Fig. 14 — Shear-Bolt Standard*

When a bottom with a safety-trip standard (Fig. 15) strikes an obstruction, a release mechanism permits the bottom to trip back over the obstacle to prevent damage to the frame or the bottom. To reset the standard, raise the plow slightly, back the tractor until the standard locks in normal plowing position, then drive forward.

JDPX6157

*Fig. 15 — Safety-Trip Standard*

Hydraulic automatic-reset standards were designed for plowing in rocky fields. Bottoms quickly swing back and rise to clear obstructions, then automatically return to working position without interrupting forward travel.

Spring automatic-reset standards are replacing hydraulic-reset plows. The newer spring-reset plows also let you plow nonstop. Bottoms swing back over obstructions and then automatically return to working position.

## Moldboard Plow Attachments

Many attachments, each serving a specific purpose, are available to improve plow performance and the quality of plowing. Among them are:

- Rolling coulters
- Disk coulters
- Trash boards
- Jointers
- Moldboard extensions
- Root cutters
- Weed hooks
- Gauge wheels

Rolling coulters improve plowing and help bury trash by:

1. Cutting trash into shorter lengths that are more easily covered.
2. Cutting through trash that might otherwise drag on the shin or moldboard and plug the plow.
3. Cutting the furrow slice vertically to provide a clear smooth furrow wall, which reduces soil pressure and wear on the share and shin.

Disk coulters have concave blades similar to a disk. These blades cut a ribbon of trash and soil from the edge of the furrow slice and turn it into the furrow bottom for better coverage.

Trash boards (Fig. 16) are mounted just above and at the leading edge of the plow moldboard, and deflect trash into the furrow bottom.

Jointers (Fig. 16) are shaped like miniature plow bottoms, and cut a small ribbon of soil just ahead of and above the share point.

Moldboard extensions (Fig. 16) attached to the rear of the moldboard provide increased turning action. They are especially helpful when plowing on hillsides or in heavy sod that turns over in ribbons.

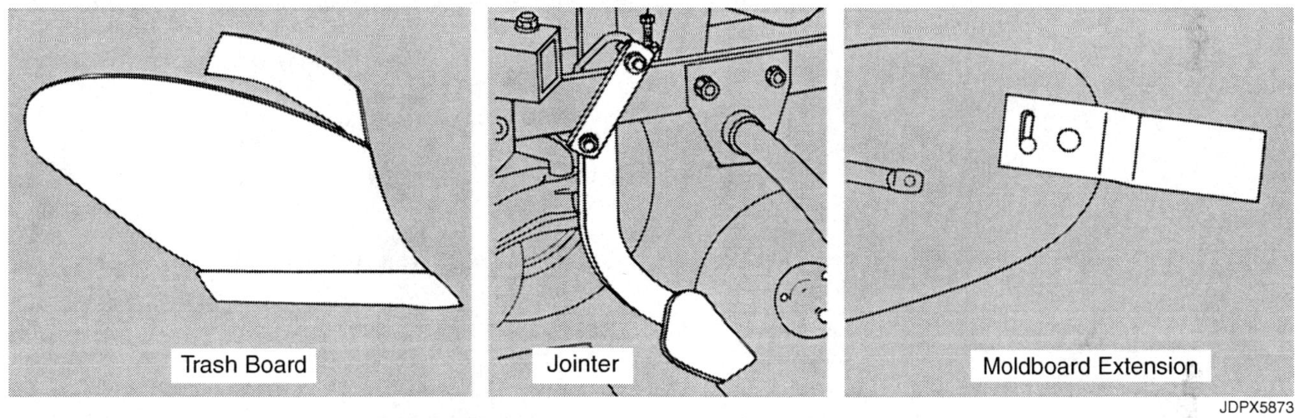


*Fig. 16 — Trash Board, Jointer, and Moldboard Extension*

## Moldboard Plow Frames

The frame is the backbone of the moldboard plow. It holds the standards in position and they, in turn, support the bottoms.

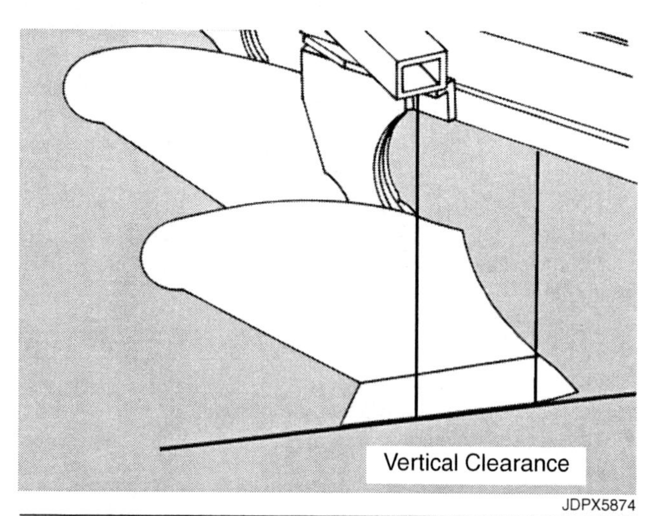


*Fig. 17 — Vertical Plow Clearance*

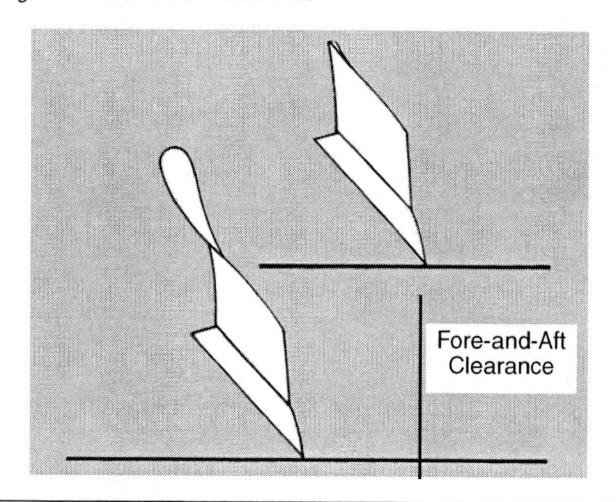


*Fig. 18 — Fore-and-Aft Plow Clearance*

Important frame dimensions are the vertical distance from the share cutting edge to the lower side of the main backbone or frame members, and the fore-and-aft distance from share point to share point (Fig. 17 and Fig).

Three basic plow-frame types are available:

- Fixed frame
- Combination
- Adjustable

Fixed-frame plows (Fig. 19) have simple construction and provide maximum strength at least cost. However, each size of plow and width of bottom requires a specific frame; for example, a six-bottom 16–inch (406 mm) plow can only be used with six 16–inch (406 mm) bottoms.

Combination-frame plows permit conversion to different sizes. For instance, a basic four-bottom plow may be converted to five- or six-bottom by adding frame extensions and bottoms. This permits matching of plow size to tractor power, traction and draft in differing soil types and conditions. Frame spacing on some models can also be changed, for instance from 14- or 16-inch to 18-inch (356 or 406 to 457 mm) bottoms.

*Fig. 19 — Fixed-Frame Plow*

Adjustable-frame plows (Fig. 20) permit varying the width of cut of each bottom to match tractor power and traction to field conditions. Some plows are adjusted manually by moving standards or shims, and others are remotely controlled from the tractor seat by the operator actuating a remote hydraulic cylinder. Plowing productivity is increased by adjusting plowing width as desired on hillsides, contours, and in finishing lands and fields.

Reducing width of cut of adjustable-frame plows reduces soil inversion and leaves more residue exposed to help control erosion. Increasing width of cut leaves larger slabs and more open spaces to catch and hold moisture.

JDPX6158

Fig. 20 — Adjustable-Frame Plow

## Types of Plows

Plow types include:

- Integral
- Semi-integral
- Drawn
- Reversible or two-way

### Integral Plows

Integral plows (Fig. 21) are attached to the tractor 3-point hitch or implement quick coupler, and the entire plow is carried by the tractor during transport. The tractor lower draft links are attached to the plow-hitch crossbar, and the upper link to the plow mast. Use of a quick coupler permits hitching by backing to the plow, lifting, and then latching the coupler automatically or manually, depending on the design.

Integral plows are limited in size, due to tractor hydraulic-lift capacity and front-end stability, though models are available with up to six bottoms. The tractor-plow combination has excellent maneuverability for short, quick turns at field ends, and easy backing into corners or plowing small, irregular fields. Integral plows require no transport wheels or axles, so they cost less than semi-integral or drawn models.

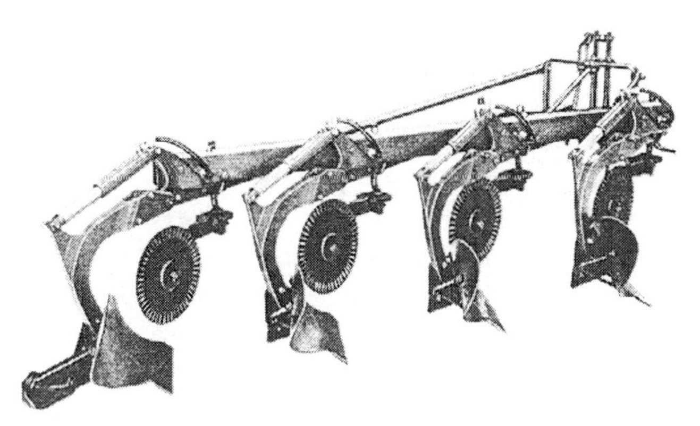

JDPX6190

Fig. 21 — Integral Moldboard Plow

### Semi-Integral Plows

Semi-integral plows (Fig. 22) are attached to hitch links of tractors with lower-link draft sensing. On double-pivot hitch plows, the upper hitch link is used to stabilize the plow on turns. The upper hitch link is also required with an implement quick coupler or if the tractor has top-link draft sensing. For top-link sensing, an A-frame is added, which places the plow hitch pins below the tractor draft links. This provides leverage on the A-frame for proper activation of the tractor weight-transfer system.

The front end of a semi-integral plow is carried and controlled by the tractor linkage, and the rear of the plow rides on a furrow transport wheel. This rear wheel is steered by a rod from the front hitch point or through a closed circuit hydraulic system between the plow-hitch crossbar and the tailwheel. When the tractor turns, oil is forced out of one end of a hydraulic cylinder mounted on the crossbar and extends the tailwheel cylinder and steers the plow. This lets the plow follow the tractor closely on turns.

A remote hydraulic cylinder raises and lowers the rear of semi-integral plows. This independent raising and lowering of front and rear permits more uniform headlands, particularly with larger plows.

Fig. 22 — Semi-Integral Moldboard Plow

### Drawn

The drawn or pull-type plow (Fig. 23) is a complete unit in itself attached to the tractor drawbar. Most current models are raised for transport or lowered for plowing by remote hydraulic cylinders. Older models have a clutch-type lift on one wheel, activated by the tractor driver with a trip-rope.

Drawn plows have front and rear furrow wheels and one land wheel, which transport the plow and provide accurate control of plowing depth.

The rear wheel usually is held rigid during plowing, but is allowed to caster for easier turning when the plow is raised. However, many larger plows have steerable front and rear furrow wheels (similar to semi-integral plows with on-land hitch) for better maneuverability and sharper turning.

On most large drawn plows, the hitch may be adjusted or interchanged to permit on-land or in-furrow operation of the right rear tractor wheel.

Pulling two plows with a tandem hitch helps utilize full power of large wheel or crawler tractors. Two smaller plows also provide greater flexibility for plowing uneven ground. compared to one very large plow. When two plows are tandem-hitched, all tractor wheels operate on land.

Some newer large plows have a flexible frame instead of using tandem plows to permit uniform plowing over uneven surfaces.

JDPX6192

*Fig. 23 — Drawn Moldboard Plow*

### Reversible or Two-Way

Traditional rollover two-way plows are equipped with two sets of bottoms, right- and left-hand, which are alternated (rolled over) at each end of the field (Fig. 24) so all furrows are turned in the same direction. They are used primarily in irrigated land where dead furrows and back furrows would impede water flow. They are also used to reduce travel time when plowing point rows in contoured fields, and to throw all furrows either uphill (preferred) or downhill, as desired.

Another model (Fig. 25) is equipped with only one set of bottoms that are hydraulically reversed (pivoted) at the end of the field. With only one set of bottoms, the plow requires less hitch lift capacity and ballast in front of the tractor as well as reduced cost for bottoms.

JDPX6193

*Fig. 24 — Integral Two-Way Moldboard Plow*

### Hitching In-Furrow or On-Land?

The operator must consider several factors before deciding whether to use in-furrow or on-land hitching. Here are the advantages of each method.

### Advantages of In-Furrow Hitching

1. Draft is reduced because soil pressure against the landside is reduced.
2. Driving is easier; the furrow wall serves as a guide, particularly for less-experienced operators.
3. It's easier to maintain uniform width of cut. Tractor wheel tread largely controls width of cut of the front bottom, so plowing is more uniform and level.
4. Weight transfer from integral and semi-integral plows is better. The front of the plow is controlled by the tractor, so a change in plow draft automatically results in weight transfer from the plow and tractor front wheels to the tractor drive wheels for more traction. Traction can be increased even more by maintaining proper vertical adjustment of the hitch.

JDPX6085

*Fig. 25 — Pivoting Two-Way Moldboard Plow*

### Advantages of On-Land Hitching

1. Reduces soil compaction. Taking the tractor wheel out of the furrow prevents compacting the furrow bottom.
2. Improves tractor pull. Equalizing weight on tractor drive wheels through level operation provides equal traction for both wheels and reduces slippage.
3. Permits use of dual wheels. In some conditions duals provide better traction and flotation and less compaction than single wheels.
4. Improves operator comfort. Keeping all wheels on-land provides a more-level sitting position. This may be somewhat offset by the need for more-accurate driving to maintain proper width of cut of the front bottom.
5. Increases tire life. With all tractor wheels on land, sidewall wear and scuffing is reduced. There's less slippage for tires that have equal footing and weight.

## Principles of Moldboard-Plow Operation

Principles of plow hitching and operation are the same, regardless of plow type.

Understanding the "hows" and "whys" of correct plow operation makes it easier to get top plow performance, and to recognize and correct problems.

Principles to be discussed include:

- Center of load
- Center of pull
- Line of draft

#### Center of Load

The theoretical center of load for a single plow bottom is assumed to be located one-quarter of the bottom width from the landside, and vertically at half the plowing depth. Thus the center of load for one 16-inch (41 cm) bottom, plowing 8 inches (20 cm) deep, is 4 inches (10 cm) from the landside and 4 inches (10 cm) above the cutting edge of the share.

Center of load for a multiple-bottom plow is found by measuring one-half the total width of cut plus one-fourth of the width of one bottom from the plowed-land furrow wall. Thus on a five-bottom 16-inch (406 mm) plow, the center of load is approximately 44 inches (1118 mm) to the left of the furrow wall.

$$\frac{5 \times 16}{2} + \frac{16}{4} = 44 \text{ inches}$$

$$\left( \frac{5 \times 406}{2} + \frac{406}{4} = 1118 \text{ mm} \right)$$

However, actual location of the center of load is affected by soil conditions, bottom type, and plowing speed. Plow and hitch adjustments have been designed to permit compensation for these variables.

#### Center of Pull

Tractor horizontal center of pull is on the tractor centerline just ahead of the rear axle, and is vertically positioned according to tractor drawbar or the convergence of top and lower hitch links. With integral plows, the center of pull is at the convergence of the links (Fig. 26). Horizontal center of pull with semi-integral plows is where the two lower links converge, and for drawn plows is at the front pivot of the tractor drawbar.

### Line of Draft

For optimum tractor-plow performance, the center of load on the plow, the hitch point on the tractor, and the tractor's center of pull must all fall on a straight line of draft. This results in minimum side-draft on the tractor, and reduces wheel slippage, fuel consumption, and plow draft. It also reduces wear on plow-wheel bearings and landsides.

To perform properly, the plow must trail straight behind the tractor, and the plow frame must be parallel to the ground-line both fore-and-aft and laterally. Proper width of cut on the front bottom is determined by proper driving, tractor wheel tread, and hitch settings.

On small or very large plows, it may be impossible or impractical to adjust tractor wheel-tread for the ideal line of draft parallel to the direction of travel. So a compromise is required and the hitch must be adjusted to match wheel settings to make the line of draft as nearly parallel as possible to the line of travel.

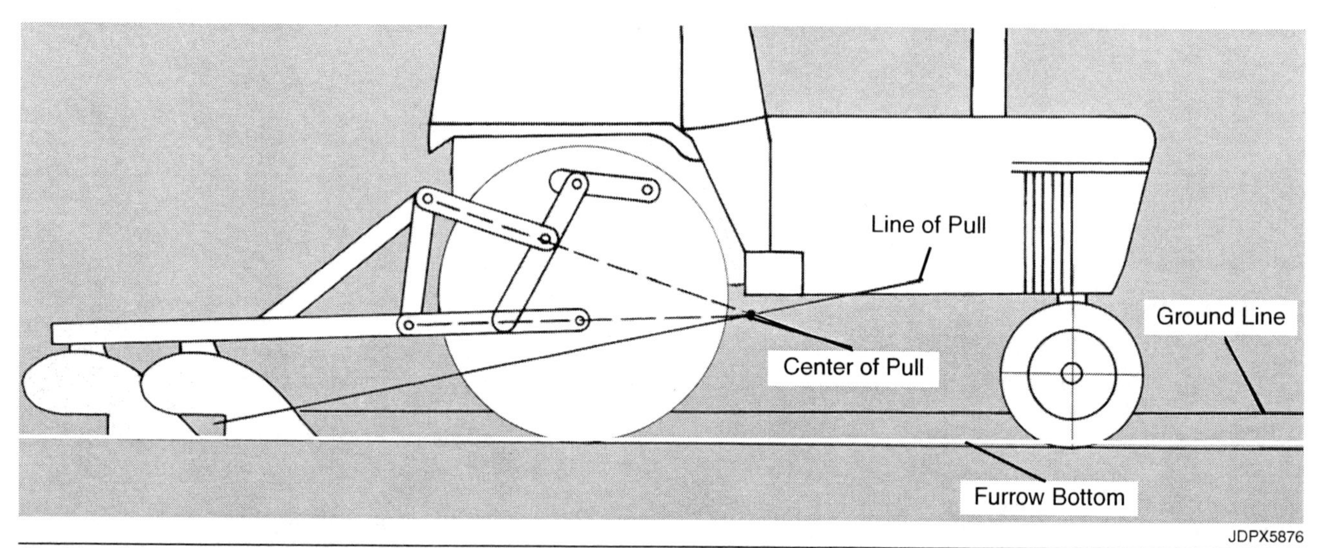


*Fig. 26 — Center of Pull With Integral Plow*

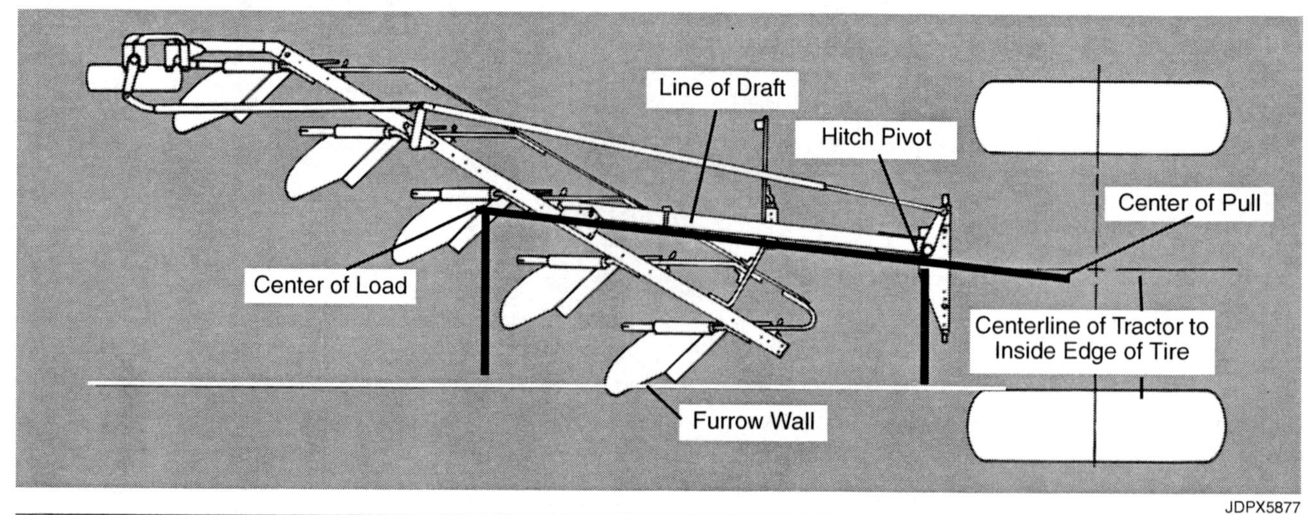


*Fig. 27 — Line-of-Draft Hitching for In-Furrow Operation of Semi-Integral Plow*

Line-of-draft setting for integral plows is established by plow designers. Tractor wheel-tread must be set according to the manufacturer's recommendations to achieve proper width of cut of the front bottom. In many cases, designers have considered such factors as common spacings for row crops, and have designed plow hitches to reduce or eliminate the need for changing tread setting between plowing and cultivation.

On semi-integral plows, proper lateral adjustment of the hitch assembly and hitch crossbar places the hitch pivot (Fig. 27) on the line of draft, between the center of load and center of pull. This adjustment is proper for in-furrow as well as on-land hitches.

With drawn plows, the hitch pin must be as close as possible on the line of draft, which passes through the center of pull and the center of load. Instructions for locating hitch points for various wheel-tread settings and either semi-integral or drawn plows are given in the operator's manual.

Accuracy of these settings may be checked by stretching a string from the center of load to the center of pull, and verifying that the hitch point falls on this line of draft.

Normal soil forces tend to rotate a plow clockwise (looking down from above). Therefore, if the hitch cannot match the true line of draft, setting the tractor center of pull to the right (toward the furrow) of the plow center of load will help offset the rotational tendency and reduce lateral pressure on tail wheel and landside.

## Basic Moldboard-Plow Adjustments

Basic principles of moldboard-plow adjustments are similar regardless of plow make. Typical adjustments that should be made include:

- Vertical hitch
- Leveling
- Width of cut
- Rolling coulters
- Disk coulters
- Trash boards
- Jointers
- Gauge wheels

Specific details and initial settings depend on design and are explained in the manufacturer's operator's manual for each plow.

## Power Required for Plowing

To realize the magnitude of work involved in moldboard plowing, and the necessity for proper plow adjustment and operation, consider the soil moved per acre (hectare). Assuming average soil density of 80 pounds per cubic foot (1280 kg/m$^3$) and 6-inch (152 mm) plowing depth, 1,306,800 pounds of soil will be moved per acre (1,464,900 kg/ha). That is 653.4 tons per acre (1465 tonnes/ha).

Viewed another way, on the average a 14-inch (356 mm) bottom travels approximately 7 miles to plow one acre (27.8 km/ha), or a seven-bottom 14-inch (356 mm) plow covers one acre for each mile traveled. (A five-bottom 410 mm plow covers one hectare for each 5 kilometers traveled.) (To determine the miles traveled per acre in plowing, or with any other implement, simply divide 99 by the operating width in inches. To determine kilometers traveled per hectare, divide 10 by the operating width in meters.)

Moldboard-plow draft ranges from about 3 to 20 pounds per square inch (2 to 14 N/cm$^2$) of furrow cross-section. Thus a 16-inch (40 cm) bottom plowing 8 inches (20 cm) deep has (8 x 16) a 126-square-inch (800 cm$^2$) furrow section. In a moist clay loam with 7 pounds resistance per square inch (5 N/cm$^2$), draft of such a bottom would be (7 x 128) 896 pounds (4000 N). Eight bottoms would have a total draft of (8 x 896) 7168 pounds (32,000 N).

As plowing speed increases, draft also goes up, but usually at a higher rate of speed. This poses a problem with a given tractor in choosing an economical balance between using a small plow at high speed or pulling a larger plow more slowly. However, most current tractors are built with less weight per horsepower for higher-speed operation.

It is normally best to plow as shallow as possible and still meet desired objectives of covering trash, aerating the soil, and providing suitable seedbed and rootbed. In most areas only minor yield differences have been found to favor deep plowing. This is particularly true where soil freezes below plowing depth during the winter. Shallow plowing also saves time and fuel.

Horsepower-hours per acre needed for plowing may range from 6.6 to 24.2 (12.2 to 44.6 kWh/ha), according to a North Dakota study. Soils ranged from silt to compacted clay loam that had been irrigated.

High horsepower required for moldboard plowing has recently caused some farmers to seek other tillage tools for seedbed preparation. However, use of other implements may require more chemicals for control of weeds, insects, and diseases. Generally speaking, for most soil and crop conditions, the moldboard plow remains the most consistently reliable tillage tool for clean-tilled row-crop seedbed preparation.

## Tractor Load-and-Depth and Draft-Control Systems

Tractor draft control maintains constant draft load on the tractor. If draft increases, draft sensing in either the tractor lower links or the top link activates the lift-control valve. Then the hitch automatically rises just enough to reduce the draft load to the preset level. As the plow or other implement is lifted, weight is transferred to tractor rear wheels for added traction to pull the increased draft load.

The lifting action on the implement also tends to lift the tractor front end, and additional weight is shifted to tractor drive wheels, providing still more traction. As draft decreases, the hitch automatically lowers until the chosen draft load is again obtained.

Tractor load-and-depth control combines the benefits of draft control, as described above, with depth or position control, which holds the hitch at a fixed height. Thus by changing the control setting, the desired system response can be obtained, from entirely draft-responsive to controlled hitch height. This reduces variations in working depth caused by soil changes.

Most tractors with top-link draft sensing respond to changes in tension and compression of the top link as implement load varies. Thus increased suction on the plow activates the control system just as an increase in soil resistance tends to rotate a smaller plow about the crossbar and push on the top link. (See John Deere's FMO Manual, TRACTORS, for a more detailed explanation of hitch-control system functioning.)

Some sacrifice must be made in uniformity of plowing depth with draft or load-and-depth control, but extreme changes are unlikely except in cases of major soil change—clay to sand, for instance. For such cases a gauge wheel on the plow will help maintain reasonably uniform plowing depth. Occasionally, if soil conditions are extremely variable, changes in load and-depth setting may be required to hold the desired plowing depth.

## Tractor Preparation for Plowing

The tractor must be matched to the plow from the standpoints of drawbar pull, hydraulic lift capacity, and tractor stability. Some preparations depend on tractor and plow design, and details are explained in each operator's manual. However, many preparations apply to all tractors and plowing situations, and should be covered before going into the field. Chapter 4 is for details on tractor preparation.

### Heavy-Duty Disks

The concept of using disk blades for tillage originated in Japan shortly after the U.S. Civil War. Factory production of disks (disk harrows) began in the United States about 1880. The disk was one of the most widely used tillage tools through the 1960s ("Selected Secondary Tillage Tools" on page 1-4). Its popularity decreased with the advent of conservation tillage, but it is still used in nearly every kind of soil condition. Heavy-duty disks (Fig. 28) are used for primary tillage such as tilling unworked soil, cutting and mixing heavy crop residue and mulching stubble. They are often used in fields with cornstalks or other residue, ahead of plowing (Fig. 29). This loosens the surface, cuts residue, and mixes it in the soil. In clean tillage systems, this operation provides better residue coverage when the land is later plowed, more soil for better soil-residue contact, and faster decomposition of residue. Lighter-duty disks are used in a variety of ways in secondary tillage operations.

JDPX6086

*Fig. 28 — Heavy-duty Disks are Used for Primary Tillage*

### Disk Types and Sizes

Variations in disk construction are common as manufacturers seek to meet specific needs of farmers. However, most disks fall into two distinct classes:

- Integral or tractor-mounted
- Drawn—with or without transport wheels

JDPX6087

*Fig. 29 — Heavy-Duty Disks Chop and Mix Residue Into the Soil*

Within each of these two classes are four primary types:

- Single-action
- Double-action, tandem, or double-offset
- Offset
- Plowing disks

Integral disks are attached to the tractor 3-point hitch and are very maneuverable for turning and transport. Tractor load-and-depth control is normally used to regulate working depth, which also is controlled by disk angle.

Drawn disks are attached to the tractor drawbar and are available in all four primary types.

Single-action disks have two gangs of disks placed end-to-end, which throw soil in opposite directions (Fig. 30). They were quite common when power was limited to horses and small tractors, but now are primarily used for splitting beds, ridges, or irrigation borders, and similar specialized tasks.

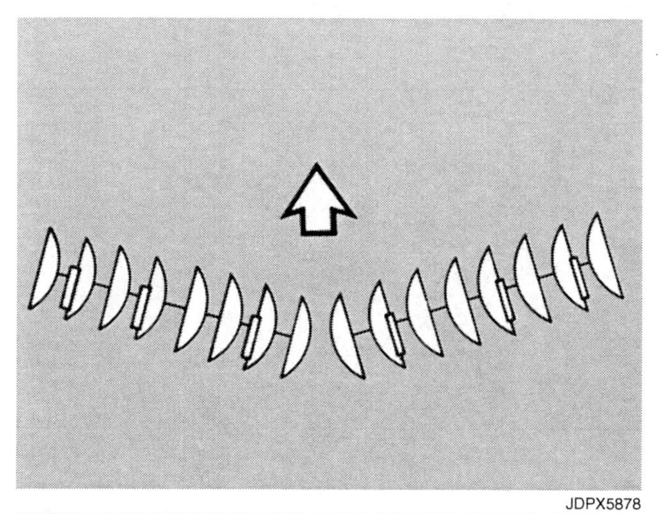


*Fig. 30 — Single-Action Disks Have Two Opposed Gangs*

Double-action, tandem, or double-offset disks have two opposed front gangs, like single-action harrows, plus two opposed rear gangs, which pull soil back toward the center of the implement (Fig. 31). Thus the soil is tilled twice with each pass and is left more nearly level, compared to single-action disking. The small furrow left on each side by the outside rear blade may be reduced or leveled somewhat by various attachments.

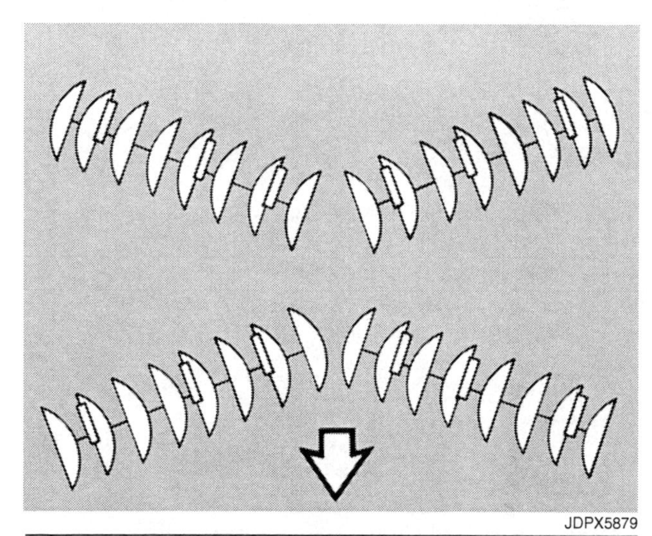


*Fig. 31 — Double-Action Disks Have Two Pairs of Opposed Gangs*

In most cases, tandem disks more than 14 feet (4.3 m) wide are designed to permit folding outer ends of each gang over the center section to reduce width for transport (Fig. 32). Some disks may be operated with wings folded to improve penetration by increasing weight applied to each cutting blade. Care must be used when disking with wings folded, because gang components usually are not designed for this extra weight.

*Fig. 32 — Tandem Disk With Wings Folded for Transport*

Offset disks have a front gang moving soil in one direction and a rear gang turning soil the opposite way (Fig. 33). Due to action of soil forces on the gangs, the hitch point and line of pull of an offset disk is considerably to one side of the center of the tilled strip. Hence the name, offset.

JDPX6089

*Fig. 33 — Front and Rear Gangs on Offset Disk Move Soil in Opposite Directions*

Plowing disks (Fig. 34) may be either tandem or offset units specifically designed for tough ground and heavy trash. Built for the most adverse disking conditions, they have heavier frames, larger blades, wider blade spacing, and greater overall strength than other disks.

JDPX6196

*Fig. 34 — Tandem Plowing Disk in Heavy Trash*

Where the deep plowing and complete trash coverage provided by moldboard plows is not desired or needed, plowing disks may be used for primary tillage. These heavy-duty disks make an acceptable seedbed in one or two passes by mixing and mulching trash into the top 6 to 8 inches (150 to 200 mm) of soil, and even deeper in lighter soils.

### Blade Types

Most disk blades are shaped like portions of a hollow sphere. The spherical radius may vary, even for blades of the same diameter, resulting in flatter or deeper blades.

Blades may be sharpened on the convex or outer side (outside bevel), or on the concave side (inside bevel). Sharpening the concave side increases penetration in hard soil, while outside beveled blades perform well in general disking conditions and in rocks.

### Cone-Shaped Blades

Cone-shaped disk blades are available for many disks and in a variety of sizes. These blades (Table 1) appear to be cut from a cone of approximately 25 degrees. The distance between working surfaces of adjacent cone disks is equal, top to bottom, which permits easier movement of soil between blades, reduces soil packing, and improves penetration. Spherical blades curve in slightly different arcs as soil passes between them, which tends to compress the soil.

JDPX6229

*Fig. 35 — Cone Disk Blades*

Cone-shaped blades are generally more aggressive than spherical blades, but may have more tendency to plug in certain sticky-soil conditions.

Notched disk blades (Fig. 36) penetrate better than plain blades in hard soil because of the reduced contact area around the outer edge. Both cone and spherical blades are available with notched edges.

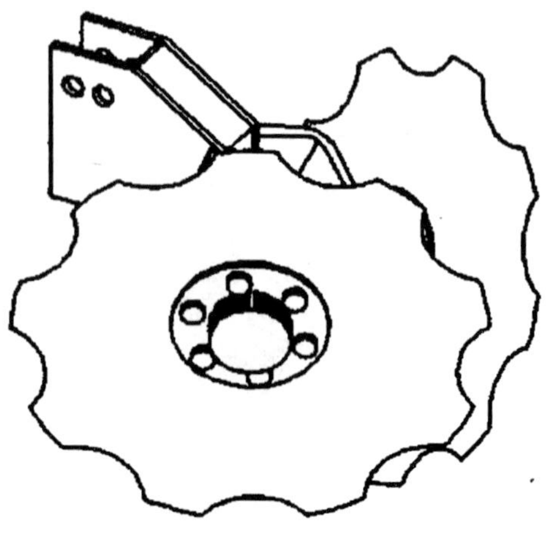

JDPX6195

*Fig. 36 — Notched Disk Blades*

### Blade Sizes

Disk blade diameters range from 16 inches (400 mm) for small, light-duty models, to 32 inches (813 mm) for some heavy offset and plowing types. Each size has its place in today's tillage operations (Table 1).

Small blades, equally weighted, will penetrate better in hard soil than larger-diameter blades because of the reduced blade-soil contact area. But larger blades cut trash better than small blades because of the angle between soil surface and cutting edge when working at equal depths.

Thickness of disk blades varies from about 1/8 inch to 3/8 inch (3 to 9.5 mm) to cover all needs from light seedbed preparation to heavy primary tillage in extremely adverse conditions.

The choice of blade thickness and diameter will depend on these factors:

- Disk type, size, and weight
- Application—primary or secondary tillage
- Soil type and moisture—dry and hard, soft, sticky
- Expected operating depth
- Type and amount of trash to cut
- Stones, stumps, or other obstructions in the soil

Blade wear-rate and impact resistance depend on type and hardness of steel used. Hard-faced blades have the best wear resistance in abrasive soils.

*Table 1 — Features of Disks Match Any Tillage Need*

| TYPE OF WORK | Blade Type | Blade Size Inches | Blade Size mm | Blade Thickness | Blade Spacing Inches | Blade Spacing mm | Typical Weight per Blade lb | Typical Weight per Blade kg |
|---|---|---|---|---|---|---|---|---|
| Final seedbed preparation | Plain | 16–22 | 400–560 | Min. | 7–9 | 180–230 | Light (30–110) | 13–50 |
| General disking | Notched | 20–24 | 500–600 | Medium | 80–100 | 200–250 | Medium (80–160) | 35–75 |
| Stalk cutting | Plain | | | | | | | |
| Chemical incorporation | Notched front | 16–20 | 400–500 | Medium | 7–9 | 180–230 | Light to Medium | |
| Seedbed preparation | Plain rear | | | | | | | |
| Heavy trash and seedbed preparation without plowing | Notched or plain | 24–26 | 600–660 | Heavy | 9–12 | 230–300 | Heavy (150–260) | 70–120 |
| Very hard soil, very heavy trash, light brush | Notched or plain | 26–32 | 660–810 | Maximum | 12–14 | 300–360 | Maximum (200 or more) | 90 |

### Blade Spacing

Selection of blade spacing is based on desired results, because the spacing between disks (Fig. 37) is directly related to tillage objectives. Spacings range from 7 inches (178 mm), for lighter disks designed for fine-finish seedbed work, to almost 15 inches (381 mm) for deep-working, heavy-duty offset harrows with very large blades. The many in-between spacings permit matching of harrow design and operating goals.

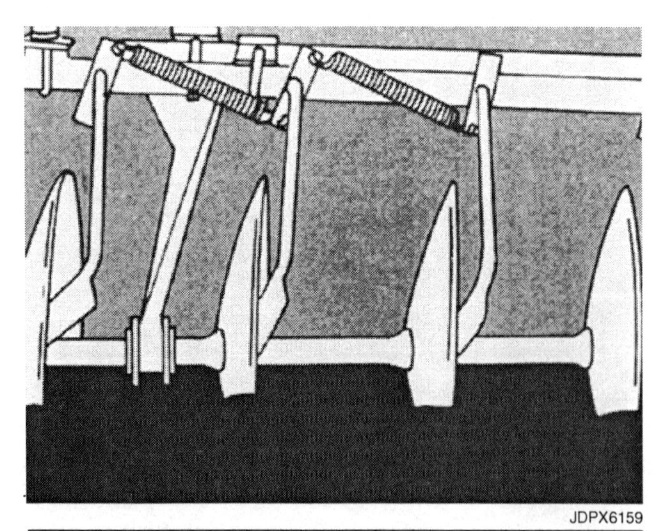


*Fig. 37 — Blade Spacing Depends on Tillage Objectives*

Spacing selection must include consideration of soil and trash conditions, size of blades, expected working depth, and intended use (primary or secondary tillage). Some compromises may be required if the same disk is to be used for primary tillage and seedbed preparation, or for cutting heavy trash and working plowed ground. The following guide can help in making the blade-spacing choice.

- 7-inch (178 mm) range—seedbed preparation in plowed soil, with relatively little trash cutting or extremely hard ground to work
- 9-inch (228 mm) range—general disking, chopping cornstalks and other trash, incorporating chemicals, and preparing seedbeds
- 11-inch (279 mm) range—deep disking in heavy trash and hard soil, seedbed preparation if soil pulverizes easily; otherwise, should be followed by lighter disk or field cultivator
- 15-inch (381 mm) range—extra—deep work with very large blades in extreme soil and trash conditions; used only for primary tillage in very severe conditions

## Principles of Operation

Each disk gang consists of a set of concave blades mounted on a common shaft or gang bolt. When the gang is operated at or close to a right angle from the direction of travel, blades tend to roll over the ground like wheels and do very little cutting. As the angle increases, disk rotation slows down, penetration increases for a given amount of disk weight, and blades scoop and roll soil as they rotate. More soil is turned and trash coverage improves as angle increases. Soil pulverization is also increased, particularly at high speeds, up to 7 mph (11 km/h).

### Gang Angle

Normal disk gang cutting angle ranges from 10 degrees to 25 degrees from a line perpendicular to the direction of travel on single-action and tandem disks, but may be up to 50 degrees on some offset disks. Increasing gang angle increases disk penetration, trash cutting and coverage, and power requirement.

Front gangs of tandem and offset disks perform the most tillage and cutting.

### Disk Penetration

Uniform, full-width penetration is the most important factor in disk operation. Satisfactory penetration requires ample strength for field conditions, proper weight distribution, and careful matching of features to desired results.

Some factors influencing penetration are controlled by disk design and selection while others are not.

Penetration factors dependent upon disk design:

- Angle of gangs—increasing angle improves penetration
- Total harrow weight and weight per blade—increasing weight improves penetration; increasing blade spacing increases weight per blade
- Blade diameter—small blades penetrate better; larger blades work deeper in soft soils
- Blade sharpness—sharp blades cut soil and trash better; inside-beveled blades penetrate better in hard soil
- Hitch angle (horizontal is best)—pulling up on hitch reduces penetration; pulling down on hitch cuts tractor traction
- Soil type—sand, loam, gumbo, clay, peat
- Soil moisture—wet, moist, dry, sun-baked
- Soil hardness—loose, firm, hard, plowed, unplowed
- Amount and kind of trash—standing or shredded corn-stalks, rice straw, bean stubble

Maximum penetration is obtained in difficult conditions by using maximum gang angle, maximum weight per blade, and with wheeled disks, raising wheels off the ground for added weight.

### Disk Leveling

The disk must be leveled laterally and fore-and-aft for uniform penetration and satisfactory performance. Integral and wheeled models are easier to level than drawn models without wheels.

If integral or wheeled disks are operated with their frames tipped lower in front, the outer ends of front gangs (or the concave end of offset disks) will penetrate deeper compared to the inner end, and soil will be worked unevenly. This problem becomes worse as gang angle increases. Rear gangs also penetrate unevenly if the harrow is not leveled fore-and-aft.

### Soil Leveling

Disks should leave soil level, but not necessarily smooth. That is, soil should be free of ridges or valleys after disking, but may deliberately be left rough to control erosion.

Leveling is easily done with offset disks by operating in lands and always filling the furrow left by the previous pass. Proper adjustment and careful driving can provide excellent leveling.

If a ridge is left in the center by a tandem disk, increasing angle of front gangs and decreasing rear-gang angle will help reduce the ridge. If a valley or low spot is left in the center, reduce front-gang angle; increase rear-gang angle; relevel the disk, or increase speed up to 6 mph (10 km/h) or the recommended limit for the disk being used.

Ridges at the outer ends of the disk can be leveled by moving rear gangs out to pull in more soil, or by reducing operating speed. Reducing the angle of front gangs also reduces the amount of soil thrown out for rear gangs to move back.

Furrow-filler blades (Fig. 38) may be added to the outer ends of the rear gangs on most disks to fill and level the furrow left by the outer blades when operating in loose soil.

## Offset Disk Procedure in Field Operation

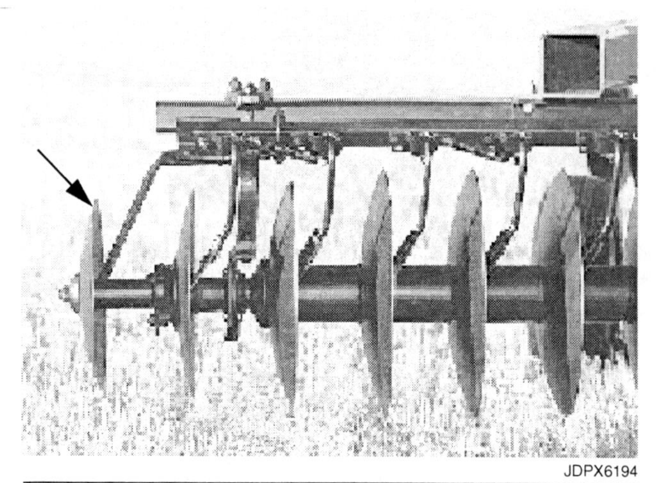

JDPX6194

*Fig. 38 — Furrow Filler Blades on Rear Gangs Help Level the Surface*

In open-field work, an offset disk is used so each pass fills the furrow left by the preceding pass, leaving the field level and free of furrows. Each pass with a right-hand offset disk is made on the right side of the previous pass. Left-hand offsets operate to the left of the last pass.

The left-end blade of the front gang of right-hand offset disks runs in the furrow to eliminate overlap and to use the full harrow width for tillage.

Furrow filling is affected by travel speed, soil conditions, disking depth, blade type, and gang angle, each of which may require different adjustments.

The right rear blade, when disking deep, may leave a furrow and small ridge (Fig. 39) These may be minimized by adding a furrow-filler blade to the end of the rear-gang axle to pull loose soil into the furrow. Most of the loose soil that was thrown onto unworked ground at the right front blade is pulled back on the next round to help fill the furrow.

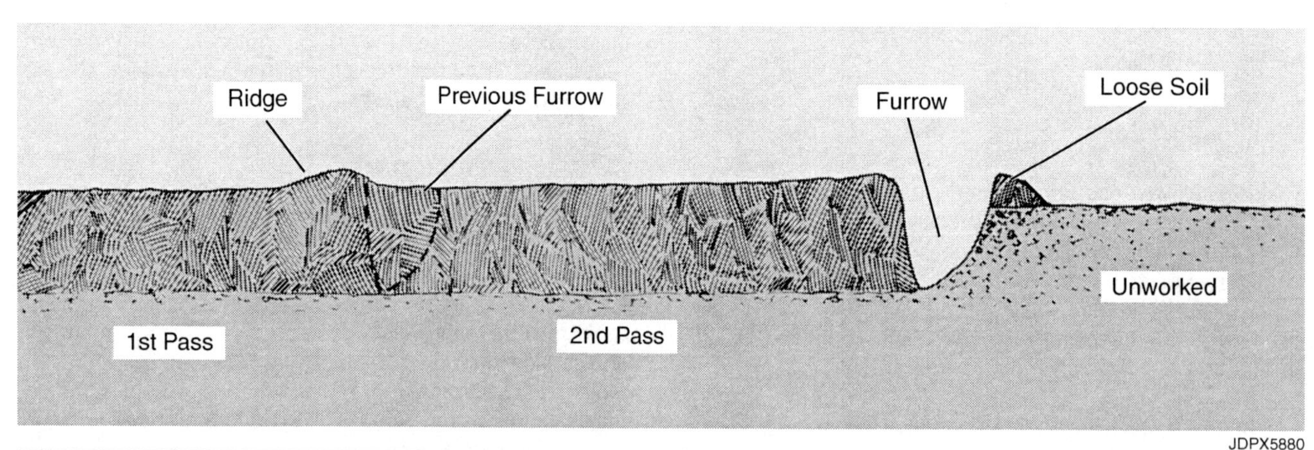


JDPX5880

*Fig. 39 — Furrow from First Pass with Offset is Filled on Second Pass*

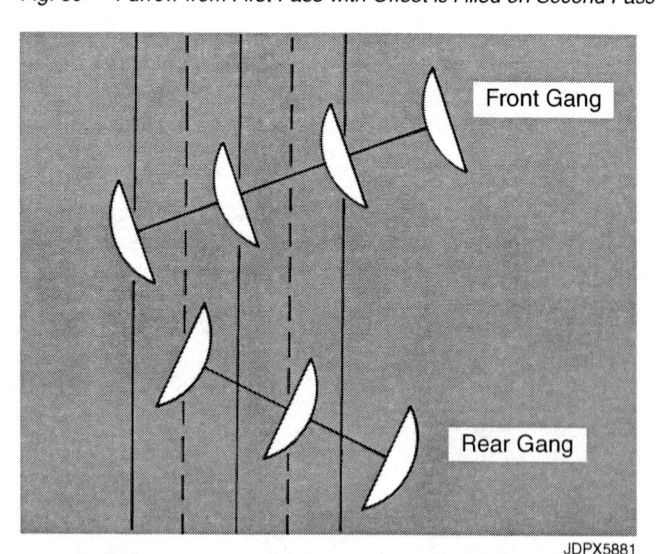


JDPX5881

*Fig. 40 — Rear Blades Should Cut Halfway Between Front Blades*

To leave most of the soil for filling the furrow, adjust the hitch as far to the left as practical. This turns the entire disk clockwise, shifts the front gang to the right and the rear gang to the left.

This reduces side draft and increases cutting angle of the rear gang so that sufficient soil is moved to fill the previous furrow.

After adjusting the hitch, operate the disk for a short distance. If necessary, shift the rear gang laterally so the left rear blade divides the space between the first two left front blades (Fig. 40). Additional lateral gang movements may be needed, depending on soil conditions, to obtain desired leveling.

**IMPORTANT: Whenever furrow size or amount of loose soil are changed, two passes must be made with the disk to be able to see the entire effect of the changes.**

JDPX6197

*Fig. 41 — Integral Seven-Bottom Disk Plow*

### Disk Plows

Disk plows are used for primary tillage, and the work performed is similar to that of the moldboard plow. They consist of a series of individually mounted, frame-supported, concave, rotating disks, with working depth of the disks controlled by one or more wheels, or the tractor hydraulic systems (Fig. 41).

Disk plows are best suited for such conditions as:

- Hard, dry soils where a moldboard has difficulty penetrating
- Sticky soils (waxy muck and gumbo) where a moldboard will not scour
- Hardpan and highly abrasive soils where the cost of moldboard plow-bottom wear would be prohibitive
- Soils containing heavy roots
- Loose-type soils, such as peat land
- Soils where deep plowing, to 12 to 16 inches (305 to 406 mm), is desired

This better performance in difficult conditions costs an approximate 10 percent increase in draft per square inch (soil cross-section turned) compared to moldboard plows. Because of this high draft, time spent ensuring top tractor performance can reduce fuel cost, save time, and result in better all-around operation.

JDPX6198

*Fig. 42 — Rotary Tillers Are Used for Many Tillage Operations*

### Rotary Tillers

Once-over seedbed preparation and reduced draft are among the most frequently mentioned reasons for using rotary tillage (Fig. 42). By applying engine power to the soil through the PTO rather than tractive force through tires, less power is lost, and tractor weight and soil compaction are reduced. High maintenance costs and power requirements relative to other tillage tools have tended to restrict their popularity. However, rotary tillers are used to:

- Shred stalks and mix them with soil
- Replace the plow, disk, and harrow
- Cultivate row crops
- Renovate pastures
- Reclaim wasteland
- Till orchards and vineyards
- Landscape
- Strip-till while planting

## Non-Inversion Primary Tillage

Non-inversion primary tillage, unlike inversion tillage, is intended to disturb the surface as little as possible and to leave it rough and covered with most or all of the residue present before tillage. This section will discuss the most popular tools now used for non-inversion primary tillage.

JDPX6190

*Fig. 43 — Wide Chisel Plows Can Work Hundreds of Acres a Day*

## Chisel Plows

The basic function of a chisel plow has changed little from that of the forked stick pulled through the soil by primitive man thousands of years ago. Alloy steels have replaced the wood and tractors have replaced animal and human muscle power, but the purpose still is to stir and aerate the soil with little inversion. Using today's wider chisel plows that can till at high speeds, one person can till hundreds of acres per day (Fig. 43).

## Chisel Plows vs. Field Cultivators

Although chisel plows are referred to in some areas as field cultivators, we are classifying them here as distinct machines. Chisel plows have heavier construction and are basically used for primary tillage. Field cultivators are used principally for secondary tillage, weed control, and seedbed preparation. Field cultivators are much lighter in construction than chisel plows, and are designed for shallower operation.

Modern chisel plows normally have two or more rows of curved spring-steel shanks attached to a rugged box-steel frame. The shanks are arranged in staggered rows (Fig. 44) to permit better trash flow and laterally balance the draft load of the machine.

JDPX6091

*Fig. 44 — Shanks Are Arranged in Staggered Rows*

## Chisel Plows vs. Moldboard Plows

Draft of a chisel plow is perhaps half that of a moldboard plow per foot of width—both working the same depth. Therefore, chiseling is faster and more economical than moldboard plowing where complete trash coverage is not required. However, if soil is chisel plowed twice to prepare a seedbed, more fuel may be used than would be needed for moldboard plowing. Chisel plows are also frequently used to break up the hardpan or plow sole formed from years of plowing at the same depth with a moldboard plow.

Because chisel plows break and shatter the soil, they perform best when soil is dry and firm. When too wet, soil is merely split open by the shank with no shattering or pulverizing. In fact, if soil is chiseled when it is too wet, large clods may be formed that are difficult or nearly impossible to break up with subsequent tillage to form a suitable seedbed.

JDPX6199

*Fig. 45 — Narrow Points and Twisted Shovels Shatter and Break Soil*

## The Versatile Chisel Plow

Chisel plows (Fig. 45) may be operated to just scratch the soil surface, or worked down to 15 inches (381 mm) or more, depending on machine design, trash conditions, and desired results. They may be equipped with narrow chisel points which dig, stir, and break the soil or with a variety of other points—from shovels for erosion control to wide sweeps for seedbed preparation and weed control (Fig. 46). Some points are hard-faced to resist wear in highly abrasive soils.

JDPX6092

*Fig. 46 — Sweeps Kill Weeds and Pulverize Surface Soil*

### Operating Speed

Operating speed depends on chisel-plow size, power available, soil conditions, and the results desired. Faster speed causes more breaking and pulverizing of the soil and more ridging, which may be desired for holding water and reducing wind erosion. However, for seedbed preparation, a somewhat slower speed will leave the soil smoother and require less additional work before planting (though work of very flat sweeps is little affected by speed).

### Chiseling Methods

Soil is normally left loose and rough after chiseling with some trash mixed under, but most crop residue remains exposed on the surface.

Some university studies indicate that approximately 25 percent of the crop residue is covered each time the soil is chiseled (Fig. 47). This, of course, depends on the amount and kind of residue and the depth of chiseling.

If the soil is to be chiseled twice before planting, it is best to work the second time diagonally to the first to break any ridges left between chisels and to prevent chisels from following the same slots in the soil.

### Stubble-Mulch Tillage

The chisel plow is an ideal tool for stubble-mulch or mulch-tillage farming; it helps prevent wind erosion and water runoff, and promotes water infiltration. If soil slopes 2 percent or more, it is advisable to chisel plow on the contour to reduce water runoff and erosion.

Chisel plows used for stubble-mulching in a summer-fallow operation are usually equipped with wide sweeps (12 to 30 inches {305 to 762 mm}) and operated just deep enough to cut off weeds with a minimum of surface disturbance.

The primary goals of a stubble-mulch system are:

- Retaining maximum surface residue to control erosion
- Encouraging infiltration and storage of the maximum amount of moisture
- Limiting surface evaporation
- Killing weeds that deplete stored water and plant nutrients

If initial residues are quite heavy, narrow chisel points may be used to work the soil from 5 to 8 inches (127 to 203 mm) deep and leave the surface rough and open.

JDPX6093

*Fig. 47 — Crop Residue Remains on Surface After Chisel Plowing*

## Chisel-Plow Types and Sizes

Most chisel plows are composed of a basic center frame, usually with one shank per foot (305 mm) of width. Rigid frame extensions 6 to 8 feet (1.8 to 2.4 m) long may be added to each end to match tractor power available and capacity needed. Folding outriggers or wings provide even greater width to match the power of big four-wheel drive and crawler tractors (Fig. 48).

JDPX6094

*Fig. 48 — Winged Chisel Plows Match Big-Tractor Power*

### Chisel-Plow Width

Chisel plows are available in sizes and styles to match almost any tractor size and field condition. Integral models are available from 5 to 20 feet (1.5 to 6 m) wide, but are limited in size by tractor power, lift capacity, and front-end stability. Drawn models start at about 10 feet (3 m) in width, with larger models introduced to match increasing tractor power.

Large plows can chisel up to 270 acres (109 ha) a day at 6 mph (10 km/h). Approximate area covered in 10 hours equals operating width in feet times speed in mph (45 x 6 = 270); or, width in meters times speed in km/h times 0.80 (80 percent field efficiency). A 14 m chisel plow traveling 10 km/h would cover approximately 112 ha per day (14 x 10 x 0.80 = 112 hectares).

Most chisel plows less than 16 feet (4.9 m) wide have rigid frames. Those more than 16 feet (4.9 m) wide are usually equipped with flexible outrigger sections. The outriggers follow ground contours (Fig. 49) and can be folded for transport.

### Chisel-Plow Clearance

Many early chisel plows offered only 18 to 20 inches (457 to 508 mm) of vertical clearance from frame members to chisel points. As crop yields and residue have increased, so has trash clearance. Many plows now have 28 to 32 inches (711 to 813 mm) of vertical clearance and as much as 3 feet (1 m) center-to-center between rows of shanks. A few frame members are 44 inches (1118 mm) apart for even greater trash clearance.

JDPX6095

*Fig. 49 — Outriggers Flex Over Uneven Ground*

To reduce overhanging weight on the rear of the tractor, some integral chisel plows have only two shank bars, with shanks spaced 24 inches (610 mm) apart on the bar. Other integral models and most drawn chisel plows have three ranks of staggered chisels that provide a foot (1 m) lateral intervals between shanks for better trash flow (Fig. 50).

Shanks may be arranged in different patterns for maximum trash flow or to work extremely hard ground. However, wheel location on some models limits choice of shank placement, particularly when using 12- or 14-inch (305 or 356 mm) sweeps.

JDPX6200

*Fig. 50 — Three-Bar Frame Provides Maximum Trash Clearance*

## Shank Types

Semi-rigid shanks are standard equipment on many current chisel plows. These shanks (Fig. 51) are clamped directly to the frame bar. They are recommended for economy, but only in soils that are free of such obstructions as rocks or stumps.

JDPX6160

*Fig. 51 — Semi-Rigid Shanks for Obstruction-Free Soil*

Various types of spring-cushion, spring-reset, and spring-trip shank mountings are available (Fig. 52). All are designed to protect the shank and frame when the point or sweep strikes an obstruction.

Shank clamps allow the shank point to lift anywhere from 6 to 14 inches (152 to 356 mm), plus any deflection of the shank, as it passes over stones or stumps. Some also allow rearward movement of the shank point before rising, which permits more uniform depth operation in extremely hard soil.

JDPX6096

*Fig. 52 — Typical Spring-Cushion Chisel-Plow Shanks*

The spring-cushioning effect of this type of mounting can produce a vibrating action in firm, dry soil, which helps break and shatter the crust.

Most current shanks are made of 1 x 2-inch (25 x 50 mm) heat-treated alloy steel for strength and durability. These shanks are naturally spring-acting because of the material and shank shape. Shanks on some heavy-duty plows are 1-1/4 x 2 inches (32 x 50 mm) or even 1-1/2 x 2 inches (38 x 50 mm).

## Soil-Engaging Tools

The wide variety of available sweeps, spikes, chisels, and shovels makes it easy to match equipment to soil conditions and desired tillage results. A representative sample of these tools is shown in (Fig. 53).

A. The regular wheatland sweep takes land polish quickly and the bottom is beveled to maintain sharpness. It kills weeds while keeping ridging to a minimum. It is available in 8- to 20-inch (203 to 508 mm) sizes.

B. High-crown sweeps lift and stir soil more than regular sweeps. They are usually available in 12- to 20-inch (305 to 508mm) sizes from 1/4 to 5/8 inch (6 to 16 mm) thick.

C. Heavy-duty, low-crown sweeps kill weeds with less soil stirring. They are available in 12- to 18-inch (305 to 457 mm) sizes.

D. Chisel sweeps break up the soil for better water absorption, and anchor stubble to retard wind erosion. Ribbing beneath the points provides added strength on some brands. The sweep shown is 6 inches (152 mm) wide.

*NOTE: Tools shown in A through D wrap around the shanks for greater strength and reduced twisting of the tool.*

E. Heavy-duty, regular sweeps are designed for maximum strength and durability. Sizes range from 6 to 14 inches (152 to 356 mm).

F. Furrow openers in 6-, 8-, 10-, 12-, 15-, and 18-inch (152, 508, 635, 762, 965 and 1168 mm) widths feature heavy points for good penetration. The soil is grooved to catch and hold water and help control wind erosion.

G. Beavertail shovels function similarly to furrow openers and are also recommended for dry conditions. The beveled bottoms retain a sharp edge. They are available in 4-, 5-, or 6-inch (100, 130, 150 mm) widths.

H\I. Reversible chisel points kill weeds and open packed or hard soil for better water penetration. The beveled edges improve cutting. They are available in sizes from 1-1/2 x 11 inches to 2 x 16 inches (38 x 279 mm to 50 x 406 mm). To double the wear, they can be reversed.

J. Double-point chisels feature forged indentations to help keep the cutting edge sharp in tough, abrasive soil. They are especially good in dry conditions. Sizes range from 2 x 16 inches (50 x 406 mm) to 2 x 18 inches (50 x 457 mm).

K. Spikes kill weeds and rip up hardpan or plow sole for better water infiltration. They are available in 2 x 12-inch to 2 x 16-inch (50 x 305 mm to 50 x 406 mm) sizes.

L. Reversible, double-point shovels kill weeds and roughen and groove the soil. This tool is an extra-wide, 4 x 14-inch (100 x 356 mm) shovel, reversible for double wear.

M. Twisted shovels feature side-throw action, somewhat like a moldboard plow, and work deep, even in heavy cover. They are good for persistent weeds in summer fallow. Also, they are reversible for longer life and available with right and left hand turn to help balance machine draft. Common sizes include 3 x 22 inches (75 x 560 mm) and 4 x 22 inches (100 x 560 mm).

Many of the most commonly used tool sizes and types are available with hard-facing for longer wear in abrasive soils.

Some farmers use chisel points or spikes for deep fall tillage. Then they switch to sweeps for fast, shallow seedbed preparation in the spring to kill more weeds and work all of the soil.

## Lifting and Folding

Integral chisel plows are lifted and carried by the tractor 3-point hitch, and these plows usually do not have any outer folding wing or outrigger sections. Wider integral plows are too wide to transport, so they are usually equipped with special transport wheels and hitch to be pulled endways.

Modern drawn chisel plows use remote hydraulic cylinders to raise and lower the plow and control operating depth. Older models have ratchet levers or hand-screw lift jacks for use with tractors without hydraulics.

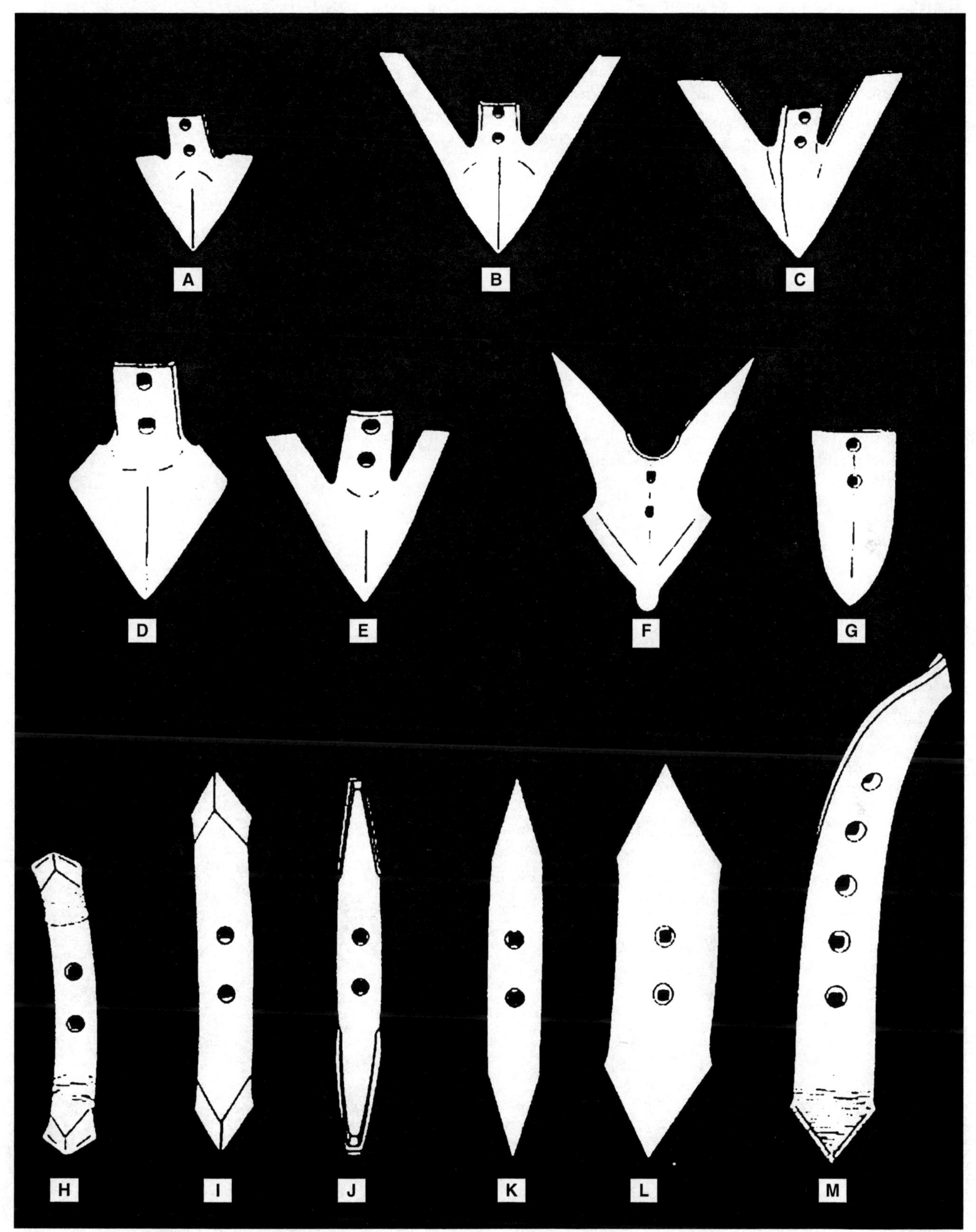


*Fig. 53 — Various Chisel-Plow Soil-Engaging Tools*

## Transport and Gauge Wheels

As an aid in depth control, integral chisel plows may be equipped with gauge wheels, usually mounted ahead of the front frame bar to avoid interference with the normal shank pattern (Fig. 54). Some gauge wheels are relocated on the frame to transport the plow endways on roads, through gates, etc.

Most rigid-frame drawn chisel plows have two transport wheels, which also gauge depth when the machine is in operation. On most plows, single wheels are also placed on each outrigger if it is more than 5 or 6 feet (1.5 or 1.8 m) long.

Some chisel plows have main frames equipped with tandem wheels on opposite sides of a walking beam (Fig. 55) to improve flotation and support and maintain more uniform working depth. When one wheel rolls over a hump or drops into a hole, its mate and the other wheels remain relatively stable. Tandem wheels stabilize the frame, distribute the weight for better shock absorption, and make transport safer. Single or tandem wheels are available for outriggers on these implements.

JDPX5882

*Fig. 54 — Gauge Wheels for integral Chisel Plows Help Control Depth*

JDPX6202

*Fig. 55 — Tandem Wheels Walk Over Uneven Ground*

## Principles of Chisel-Plow Operation

Chisel plows are designed to penetrate hard soil, shatter compacted layers, and break up large clods. The surface is left broken and open to catch and hold rainfall and resist wind erosion. Most crop residue is left on the surface, where it helps reduce evaporation and erosion.

Tests have shown that minimum draft on such tools as chisel plows and subsoilers occurs when the lift angle is 20 degrees between the face of the tool and horizontal. Shattering occurs with the least effort when the tool is applying a bitting force, rather than cutting horizontally or pushing vertically against the soil. Thus the common curved shank (Fig. 56) is ideally suited to provide optimum soil fracturing with reduced draft.

JDPX6201

*Fig. 56 — Curved Shank Does Good Work With Least Draft*

Operating a chisel plow so deeply that the upper curved portion of the shank is pressing down on the soil can increase draft unnecessarily. The alternatives are to get a machine with more clearance or work the land twice—once at a shallower depth, and then, at right angles or diagonally to the first pass, chiseling at the maximum desired depth.

Most tests indicate a moderate increase in specific draft (draft per square inch {$cm^2$} of tilled cross-section) as depth increases. It is difficult to predict the effect of increasing depth on total chisel-plow draft because of the following reasons:

- Variations in soil type
- Moisture conditions
- Shape of the sweeps and points used
- The angle at which the points penetrate the soil

Draft also increases as speed increases, but again tests show varying results depending on depth, surface area of the sweeps or points, lift angle, and soil conditions. As speed increases, so does soil shattering. Therefore, though more power is expended for fast chiseling, it may be offset later by reduced additional work required for seedbed preparation.

### Hitching the Chisel Plow

Integral chisel plows are attached to the tractor 3-point hitch in the usual manner (Fig. 57) with or without an implement quick coupler. They are leveled laterally by adjusting the length of the lift links. To provide uniform penetration of all shanks, the machine must also be leveled fore-and-aft by adjusting the tractor center link. Follow procedures and recommendations in operator's manual.

JDPX6224

*Fig. 57 — Integral Chisel Plows Are Carried on 3-Point Hitch*

Tractor load-and-depth control is often used with integral chisel plows to regulate operating depth and provide weight transfer for better traction. However, in varying soil conditions it may be desirable to provide gauge wheels to limit operating depth and reduce the need of manually changing load-and-depth control frequently. When properly adjusted, gauge wheels should carry very little weight when plowing at the desired depth in heavy soil when the load is highest. Gauge wheels carry more weight as draft is reduced in lighter soils. The tractor control system's usual reaction would call for increased operating depth to maintain uniform draft.

Drawn chisel plows are usually equipped with a rigid hitch that is attached to the tractor drawbar. Vertical hitch adjustment is provided for leveling the machine fore-and-aft to compensate for changes in operating depth, variations in drawbar height, tire size, and soil conditions.

## Field Operation

Chisel plows may be operated satisfactorily in many field patterns (Fig. 58). Work may be started at one side of the field and adjacent passes made until the field is finished. Headlands, usually about twice the width of the machine, are worked last.

JDPX6099

*Fig. 58 — One of Many Chisel-Plow Field Patterns*

Chisel plows may also be worked in lands with headlands plowed last, or operated around the field until reaching the center.

Regardless of the plowing pattern, the plow must be raised from the ground when making sharp turns. This makes steering much easier and protects shanks and the frame from heavy, twisting side forces.

### Plowing Angle

To provide better leveling and maximum loosening of row-crop stubble, such as cornstalks, it is best to cross rows at an angle of 20 to 30 degrees. This ensures that all roots are cut, spreads the ridges that were formed during cultivation, and allows for better residue clearing. If the crop was planted on ridges or furrowed for irrigation, it is usually necessary to follow the rows for the first pass. Subsequent operations are then made diagonally or at right angles for better leveling and more thorough working of the soil.

JDPX6209

*Fig. 59 — Chisel Plow With Rod-Weeder Attachment*

Adding a rod-weeder attachment to a chisel plow combines many advantages of both machines—deep tillage, weeding, packing, and mulching (Fig. 59).

JDPX6100

*Fig. 60 — Subsoilers Break Hardpan*

## Subsoilers

Subsoiling usually is done to loosen or break up impervious soil layers below the normal tillage depth to improve water infiltration, drainage, and root penetration (Fig. 60). To be effective in improving crop yields, subsoiling must meet these conditions:

1. It should be done when soil is relatively dry for maximum effect on the hard layer. If soil is wet, only a thin slot is sliced through the soil, which will likely reseal very quickly, and down-pressure of tractor weight and subsoilers can cause compaction.
2. Soil below the impervious layer must have excess water-holding capacity, or there will be no place for surface water to go, and no air in the deeper layers for plant-root growth.
3. Deeper soil must not be so acid or alkaline as to discourage root growth.
4. Tractors and heavy implements must run at least one foot (30 cm) away from subsoil slots during subsequent operations, to prevent resealing of the slot by tire compaction.

Some outstanding results have been achieved from subsoiling. Yield increases of 50 to 400 percent have been reported from subsoiling under the right soil and moisture conditions and in the right areas. But, in other cases, little effect has been observed.

Depending on subsoiler design, power available, soil conditions, and depth of hardpan, subsoiler penetration may be as deep as 24 inches (60 cm) (Fig. 60).

## Subsoiler Types and Sizes

Most subsoilers in current use are integral models with up to 13 standards for various tractor sizes and depths of penetration. Wheeled subsoilers have been common in some areas in the past and are still used for severe conditions. Increased tractor size has made it practical to operate larger integral subsoilers, which are more convenient and maneuverable.

When two or more subsoiler standards are used, they are usually attached to a box-bar or tubular toolbar that permits wide variations in spacing (Fig. 61).

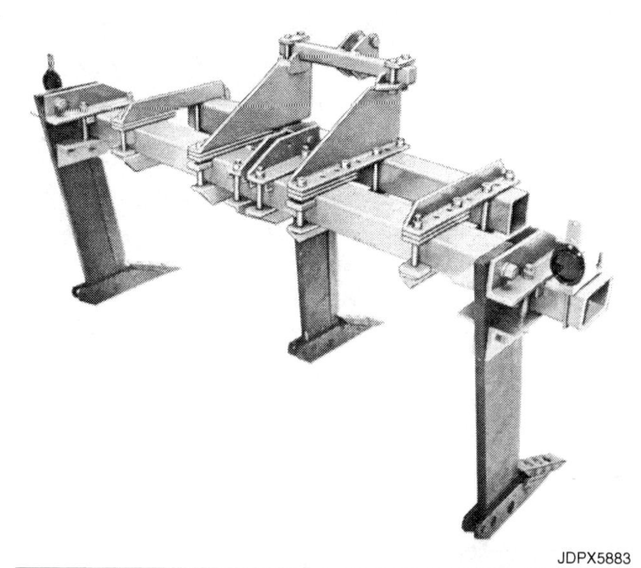

JDPX5883

*Fig. 61 — Toolbar Mounting Permits Varying Space and Number of Standards*

Many of these standards are equipped with a shear bolt or even better, a safety-trip (Fig. 62) to protect subsoiler and tractor from damage if the subsoiler strikes an obstacle.

JDPX6203

*Fig. 62 — Safety-Trips Protect Standards From Sod Obstacle*

A reversible, replaceable, tapered shin on some standards reduces draft and provides longer service and better soil cutting.

A subsoiler variation is the V-shape subsoiling chisel plow (Fig. 63), or V-ripper or V-chisel, which has from 5 to 13 standards for ripping hardpan as deep as 16 inches (400 mm). Standards may be located from about 18 to 40 inches (457 to 1016 mm) apart. Standard or optional gauge wheels control depth of penetration.

JDPX6204

*Fig. 63 — "V" Subsoiling Chisel Rips Hardpan Down to 16 inches (41 cm)*

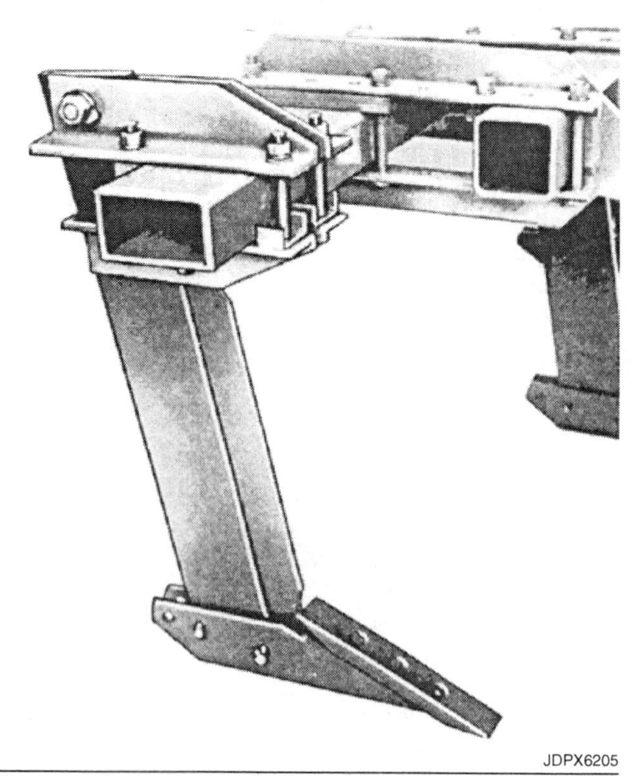

JDPX6205

*Fig. 64 — Sloping Standards Lift and Break Soil*

Slope of subsoiler standards and points affects draft and soil shattering. When standards are inclined forward (Fig. 64), they lift and break the soil much better than if they are vertical, or nearly so. Curved standards (Fig. 65) work under hardpan, lifting and shattering the soil ahead of and between standards.

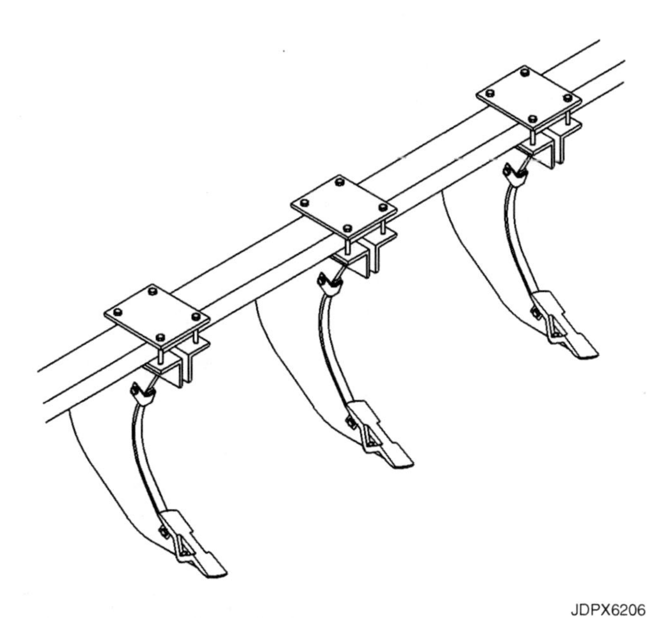


*Fig. 65 — Curved Standards Penetrate Under Soil for Better Shattering*

Staggering subsoiler standards on the toolbar (Fig. 66) provides better trash clearance and permits easier operation because the front row of standards helps break the soil for the second row.

Subsoiling chisels leave the soil surface rough, open, and loose (Fig. 63) to absorb water and reduce erosion. Deep penetration permits deeper water infiltration and increased storage for crop production.

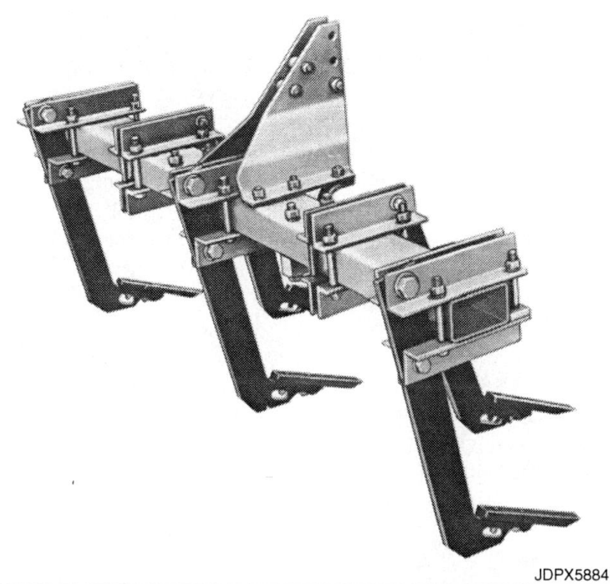


*Fig. 66 — Staggering Subsoiler Standards Reduces Draft and Improves Trash Clearance*

## Field Operation

Subsoiler operating patterns depend on the objectives. If subsoiling and bedding are a single operation, the pattern must follow the desired future row pattern. Where soil is irrigated, the same row pattern is usually followed year after year. It may be desirable to subsoil and place new beds over old row middles. This will break down the old bed, uproot crop stubble, and help cover residue.

Toolbar subsoiler spacing is adjustable and is usually 3 feet (1 m) or more. "V" subsoiling chisels may be arranged from 18 to 40 inches (457 to 1016 mm) apart on the bar. Single-standard subsoilers are usually operated at 3- to 8-foot (1 to 2.5m) intervals, depending on soil conditions and operator preference.

Operating the subsoiling chisel on a contour or across the path of prevailing winds will help reduce erosion and increase water-holding capacity.

### Mulch Tillers

Concerns for energy costs, soil erosion, water pollution, and operating costs led to adoption of conservation tillage systems by many farmers. As much as 25 percent of United States cropland is now under some form of residue management for conservation farming. Federal legislation that requires conservation on erodable soils likely will intensify adoption of conservation tillage systems.

In the 1970s, equipment manufacturers developed tillage implements from combinations of basic tillage tools that had been popular for many years. These combinations are useful for both primary and secondary tillage. Combination tillage implements were quickly adopted; by 1990, combination tools for primary tillage accounted for 10 percent of industry tillage tool shipments and the percentage continues to grow.

Combination implements have become very popular because they can perform more than one tillage operation in a single pass, thus reducing the number of trips across a field. Also, they can be equipped and adjusted to leave plant residue on the soil surface rather than burying it in the soil. Consequently, they fit very well into the trend of reduced or conservation tillage.

JDPX6101

*Fig. 67 — Mulch Tiller Is Hybrid-Disk or Coulter and Chisel Plow*

The mulch tiller (also called stubble-mulch tiller or cutter chisel plow) is a combination tillage implement that combines the disk or coulter and chisel plow into one machine typically used for primary tillage (Fig. 67). It is especially well suited for stubble mulch or deep fallow tillage in corn or small grain fields.

The front section of the mulch tiller is a disk or coulter gang that cuts and sizes residue and loosens corn or small grain stubble. To the rear, a chisel plow rips up the packed soil, breaks through old plow soles, and helps mix residue into the soil.

It's often possible to work cornstalks immediately after harvest with no prior stalk chopping or disking. One pass leaves the soil mulched, broken, and rough-ready to absorb winter moisture and resist wind and water erosion. The reduced number of trips over the field helps minimize root-restricting compaction and improves water infiltration.

Because of the cutting and mixing action of the disk gangs, the mulch tiller covers more residue than a regular chisel plow. This is particularly advantageous in heavy cornstalks where all of the residue may not be required on the surface for erosion control.

The degree of residue sizing, depth and aggressiveness of tillage, and the percentage of residue left on the soil surface can be controlled by independently regulating depth of coulter or disk and by the choice of several types of chisel plow shovels and sweeps. The result is a versatile primary tillage implement that combines operations into one pass while retaining surface residue. It fits the needs of most conservation tillage systems.

Generally, a tractor capable of pulling a six-bottom, 16-inch (400 mm) moldboard plow can pull an 11-foot (3.4 m) or larger mulch tiller at equal or slightly higher speeds. Recommended available power ranges from about 12 to 20 drawbar horsepower (9 to 15 drawbar kW) per chisel shank, depending on soil type, moisture content, working depth, and soil-engaging tools used. Thus more ground can be worked in the same period of time. There are no dead furrows or back furrows, and machine investment is roughly equal.

Tiller working depth in the fall may be as much as 8 to 14 inches (20 to 40 cm) to break through plow sole or hardpan or to increase the soil's winter moisture-holding capacity. Working depth of the machine is controlled by adjusting the depth stop on the remote hydraulic cylinder. Manual adjustment is provided to alter relative cutting depths of disk gangs and the chisel-plow section.

## Mulch-Tiller Types and Sizes

Mulch tiller capacity depends on operating width, power available, operating depth, and the number of chisel-plow shanks used (Fig. 68). On most machines, 15-inch (380 mm) shank spacing is standard, while other shanks are on 12-inch (300mm) centers, depending on design and working conditions. Different shank patterns are recommended for maximum soil ridging, for better soil leveling, and for maximum residue clearance.

Shanks must be correctly positioned on the frame to avoid uneven draft and poor performance. An equal number of shanks must be used on each side of the machine, and the location of individual shanks determined by measuring from the frame centerline.

Aggressive cutting of residue and roots is provided by large disk blades on the front gang of the typical mulch tiller. For improved leveling, the outer blade on each gang is slightly smaller in diameter. Disk spacing is usually 9 or 12 inches (229 or 304mm).

JDPX6108

*Fig. 68 — Field Capacity Depends on Machine Size and Type of Work*

A variation of the mulch tiller uses chisel-plow shanks on the rear, but has flat coulters instead of angled disk gangs in front. Widths available are approximately 6 to 26 feet (2 to 8 m). Coulters are spaced 7-1/2 or 12 inches (190 or 300 mm) apart and slice through trash and roots to permit easier penetration of the chisel plow and reduce residue plugging. Coulter operating depth is adjustable, and should be just deep enough to slice through trash. Attempting to cut too deep with the coulters, especially in hard, dry soil, may tend to lift the machine from the ground and reduce the blade-soil angle. This makes cutting less effective, because coulters tend to push residue instead of cutting it.

JDXP6225

*Fig. 69 — Disk Blade Variation of Mulch Tiller*

Other mulch tiller variations include a machine with spherical disk blades individually mounted at 12-inch (300mm) intervals across the front of a chisel plow frame. Chisel shanks are also spaced 12 inches (300mm) apart.

Another version has widely spaced disks set to throw soil outward from the center of the machine, followed by a second set of disks angled to move soil back toward the center, thus leveling the surface ahead of the chisel shanks. The nominal disk spacing is 12 inches (300mm).

JDPX9109

*Fig. 70 — Mulch Tiller Equipped With Curved Subsoiler Shank*

Two similar rows of wide-spaced disks are used on another machine equipped with curved subsoiler shanks (V-ripper) rather than chisel shanks (Fig. 70).

Rigid chisel shanks are standard on mulch tillers. They are most economical in light-soil fields free from rocks or other hidden obstructions. Where rocks are a problem or soil is average to high draft, various types of reset shanks are available that allow the point to lift up to 11 inches (28 cm) to clear stones or other obstructions. A parabolic shank, also available for some models, fractures more soil and disturbs more residue than straight shanks.

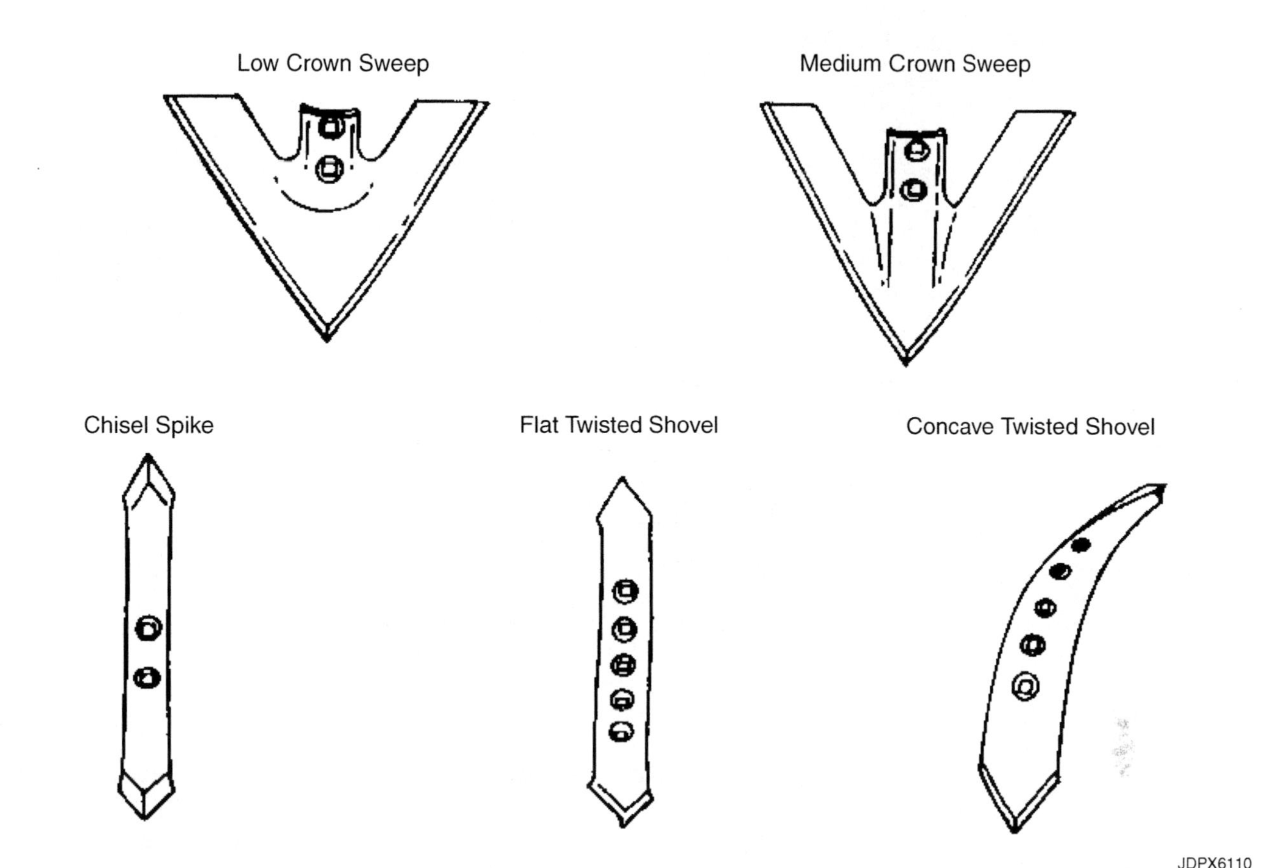


*Fig. 71 — Some Shovels and Sweeps Used on Mulch Tiller Chisel Plows*

Twisted shovels are often used on the chisel plow portion of the mulch tiller, particularly for deep fall tillage, but a range of sweeps and shovels is available for specific conditions. Some of the most commonly used types are shown in (Fig. 71).

Low and medium crown sweeps (Fig. 71) have become more popular with farmers who want to retain more residue on the soil surface. Low crown sweeps leave a maximum of residue. They are excellent for both summer fallow operations or in row-crop tillage. Because they stir the soil less, fields are left in a less ridged condition.

Medium crown sweeps perform very much like low crown versions, but are slightly more aggressive. The result is more soil stirring and slight ridging. Both low and medium crown sweeps are equally effective at shallow depths or up to 10 inches (25 cm) deep.

Chisel spikes (Fig. 71) have a beveled point to open the soil for water penetration even below depth of penetration. They are designed for deep tillage and maximum field ridging. They also leave more residue on the surface than twisted shovels.

Flat twisted shovels (Fig. 71) perform much like a mini-moldboard plow bottom. They are used for deep tillage where more residue incorporation and mixing is desired. They leave the field in a roughened, erosion-resistant condition.

Concave twisted shovels (Fig. 71) cut deep grooves in the soil and are used where maximum residue incorporation is desired.

On some models of mulch tillers, a shovel with attached wings is available for more aggressive tillage. This shovel lifts, twists, and rolls the soil rather than gently lifting it like straight shovels.

## Principles of Mulch Tiller Operation

Stubble-mulch tillers provide more complete cutting and mixing of trash and deep shattering of the soil in one operation than either disks or chisel plows alone. Residue coverage and soil pulverization are controlled by adjusting relative working depth of coulters, disk gangs, and chisel points, so the surface is left in the desired condition.

JDPX6207

*Fig. 72 — For Better Leveling, Work Diagonally Across Rows*

## Field Operation

Operating procedures for mulch tillers are similar to those for chisel plows. Always raise the machine completely out of the ground before making sharp turns or backing up to avoid shank and disk damage. Level the frame fore-and-aft and side-to-side for uniform penetration.

As with chisel plows, mulch tillers do not require any specific pattern of field operation for satisfactory performance. Two passes at different angles may be required in extremely hard soil. Better leveling of row-crop ridges and uprooting of crop stubble is obtained if the mulch tiller crosses rows diagonally at a 20- to 30-degree angle (Fig. 72). Otherwise, it may follow the row pattern, or be operated in lands, in adjacent passes across the field. Working on the contour or across slopes improves erosion control.

The mulch tiller is effective in heavy residue (Fig. 68), such as cornstalks, where it is desired to chop and mix a portion of the residue with the soil and leave enough exposed to control winter erosion. For such work, set disk gangs at maximum angle and fairly deep to provide maximum cutting. Such a setting will also help loosen plant roots, break up surface soil, and mix residue for faster decomposition.

Set coulters just deep enough to slice through residue. Coulters set too deep waste power and may push residue in front of the blades instead of cutting it. If the soil is very hard, coulters set too deep may hold the front of the machine out of the ground.

Set chisels to penetrate the root zone and break soil open for maximum entry of air and water. The breaking and stirring action of the chisels also helps anchor residue and roughens the surface to improve its erosion resistance.

If too much residue is being covered for satisfactory erosion control, reduce disk-gang angle and cutting depth if necessary. If too much residue remains exposed, it may be necessary to shred or disk stalks prior to using the mulch tiller. If residue is disked prior to using the coulter-type mulch tiller, allow the soil to settle between operations to permit better cutting by the coulters.

## Anhydrous Ammonia Applicators

Primary tillage machine versatility is provided by using anhydrous ammonia applicators behind sweeps or shovels of chisel plows. The ammonia tank can be mounted directly on the plow frame (Fig. 73) or on a trailer behind the plow.

While it is not usually considered a tillage operation, shanking in anhydrous ammonia can disturb significant amounts of surface residue. Some estimates indicate that this operation, the equivalent of one tillage pass, can bury up to half of the surface residue. Therefore, fertilizer incorporation must be considered in estimates of residue coverage when developing a conservation tillage system.

JDPX6208

*Fig. 73 — Anhydrous Ammonia Applicator*

## Summary

Primary tillage works soil to a depth of 6 inches (15 cm) or more. Some primary tillage tools completely invert the soil to bury residue. Others work the soil to the same depth, but leave the surface covered with most or all of the pre-tillage residue.

Moldboard plows, heavy-duty disks, disk plows, and rotary tillers are examples of inversion primary tillage tools.

Moldboard plows and disks are most often used in inversion tillage, but their popularity has decreased significantly with continued adoption of conservation tillage systems. The plow is one of the best tools available for burying surface residues and weed seeds. But moldboard plowing is also a very high power-consuming operation.

Heavy-duty disks continue to be used in almost every soil condition. Many farmers consider them superior to chisel plows and field cultivators for working chemicals into the soil. They can be used to good advantage before or after moldboard plowing in clean tillage systems.

With increased use of conservation tillage, non-inversion primary tillage has assumed a greater role. Chisel plows, subsoilers, and mulch tillers are most frequently used.

Chisel plows leave a rough surface covered with residue, ideal for moisture retention and reduced erosion. They accomplish little in wet soil, but are very effective when soil is dry and firm, and are at their soil-shattering best at fairly fast speeds in such conditions.

Subsoilers break up impervious layers of soil below normal tillage depth to improve root and water penetration.

The mulch tiller combines a disk or coulter and chisel plow into one machine. Mulch tillers are available in several design configurations from manufacturers. They are popular because they perform more than one tillage operation in a single pass, thus reducing the number of trips across a field. They also can be adjusted to leave plant residue on the surface rather than burying it. Consequently, they fit very well into reduced or conservation tillage systems.

# Test Yourself

## Questions

1. Define primary tillage.
2. What is inversion tillage? Name two tillage tools used in inversion tillage.
3. List at least two advantages of the moldboard plow. What is a major disadvantage of the plow?
4. Name four key parts to a plow bottom.
5. What usually limits the maximum size of integral moldboard plows?
6. What are the four primary types of plowing disks?
7. What are the benefits of disking after moldboard plowing?
8. (T/F) Use of moldboard plows and heavy-duty disks increased two-fold from 1980 to 1990.
9. (T/F) Chisel plows perform best when soil is soft and moist.
10. Why are chisel plows popular in conservation tillage systems?
11. List six types of soil-engaging tools used on chisel plows.
12. (Fill in blanks.) The mulch tiller is a combination of _________ and __________.
13. How is residue coverage and soil shattering controlled when using a mulch tiller?
14. What type of shank is used on most mulch tillers?
15. (T/F) The mulch tiller is relatively ineffective in heavy residue such as cornstalks.

# Secondary Tillage

# 7

JDPX6161

*Fig. 1 — Integral Disk*

## Introduction

Secondary tillage operations, less aggressive than primary tillage, are used to prepare fields for planting, for summer fallowing, chemical incorporation, and weed control. The degree of tillage varies with the type of tillage system. With a conservation tillage system, one pass primarily to reduce surface roughness or kill small weeds might be sufficient. A clean tillage system may require two or more passes with more than one tillage tool to provide a well-tilled surface free from residue and weeds. Light and medium disks, field cultivators, combination tillage implements, and various harrows and packers are commonly used in secondary tillage operations.

## Disks

Light to medium disks, or properly adjusted and equipped heavy-duty models, are well suited for secondary tillage. With the rapid adoption of conservation tillage systems, disks are currently used more frequently for secondary than for primary tillage.

Disks used primarily for secondary tillage are available as integral (Fig. 1) or drawn (Fig. 2) types. Operating widths vary from about 6 feet (1.8 m) to 32 feet (9.8 m) in folding models (Fig. 3). Furrow filler, tine-tooth harrow, and finishing blade accessories add to the capacity of disks to provide a finished smooth seedbed.

Chapter 6, Primary Tillage, provides a detailed description of disks and their features, including models used for secondary tillage, their types and sizes, blade characteristics and spacing, and principles of operation.

JDPX6111

*Fig. 2 — Drawn Disk*

JDPX6088

*Fig. 3 — Wide Disks Fold for Transport*

## Field Cultivators

Field cultivators are widely used across North America for seedbed preparation, weed control, stubble-mulch tillage, summer fallow, and roughing fields to increase moisture absorption and control wind and water erosion.

Field cultivators and chisel plows have similar appearance and operating characteristics, and in some areas chisel plows are known as field cultivators. But the field cultivators described in this chapter are lighter than most chisel plows and are intended for less-severe operating conditions (Fig. 4).

When soil permits, field cultivators occasionally replace chisel and moldboard plows for primary tillage as well as secondary tillage. Usually, however, they are intended for secondary tillage of previously worked soil, such as seedbed preparation in fall- or spring-plowed fields, use after stalks or stubble have been chisel plowed or disked, summer fallowing after use of a disk tiller, chisel plow, or wide-sweep plow, and similar operations.

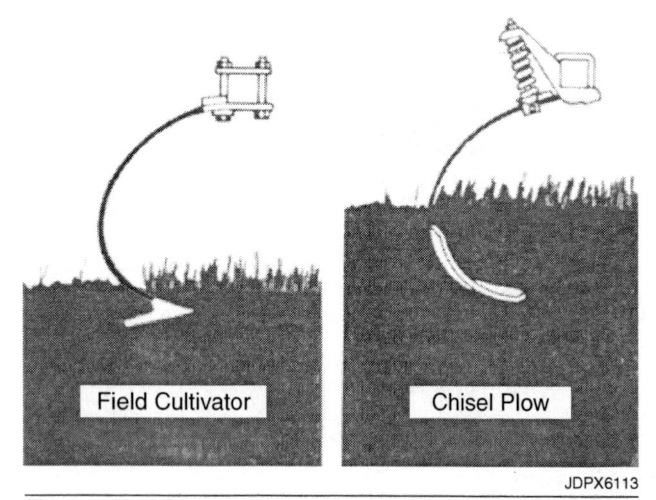


JDPX6113

*Fig. 4 — Field Cultivators Are Intended for Less-Severe Conditions*

Field cultivators leave most of the residue on top or mixed into the upper few inches of soil. When using spike points, the surface is usually left ridged, rough, and open so moisture infiltration is increased and soil blowing and runoff are reduced.

### Vertical Clearance

Vertical clearance of field cultivators is generally much better than that of spring-tooth harrows, but somewhat less than chisel plows because of the normally narrower tooth spacing, 6, 9, or 10 inches (152, 229, or 254mm) in most cases, compared to 12 inches (305 mm) or more on chisel plows. Clearance is usually 19 to 22 inches (483 to 559 mm) compared to 26 to 32 inches (660 to 813 mm) for chisel plows.

Working depth depends on soil conditions, desired tillage results, and soil-engaging tools used. Double-pointed shovels and similar narrow points may be used down to 5 inches (127 mm) or more in lighter soil with relatively little residue. When sweeps are used for weed control or seedbed preparation, depth is usually about 4 inches (102 mm). If working freshly plowed soil, particularly sod, depth must be limited to prevent dragging chunks of sod or residue to the surface.

### Types and Sizes

Integral field cultivators (Fig. 5) range from 7-1/2 to 24-1/2 feet (2.3 to 7.5 m) wide. Those wider than 15 feet (4.5 m) are equipped with folding wings to reduce transport width. Drawn models (Fig. 6) are available from about 8-1/2 to 60 feet (2.6 to 18 m) or more to match tractor power and acreage. Various sizes may be assembled by adding stub bars, extensions, or folding wings to basic integral or drawn center sections. Extremely wide cultivators may have two wing sections on each side to provide maximum flexibility and reduce overall height of folded wings (Fig. 7).

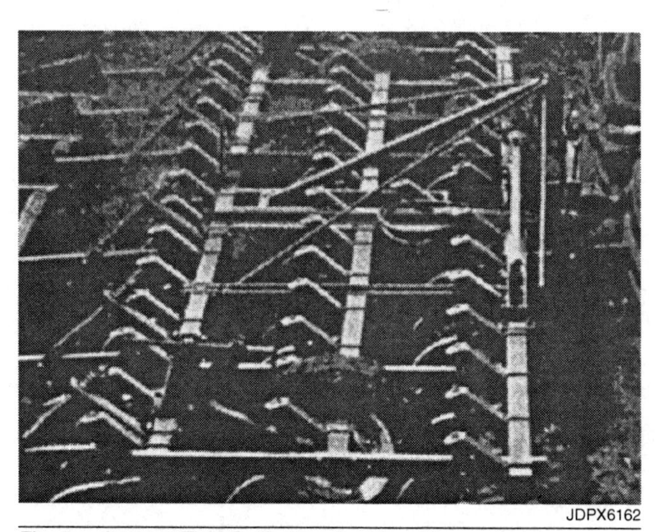

JDPX6162

*Fig. 5 — Integral Field Cultivators Are Very Maneuverable*

JDPX6114

*Fig. 6 — Drawn Field Cultivator*

Integral cultivators may be attached directly to the tractor 3-point hitch, or to an implement quick coupler for pick-up-and-go operation.

JDPX6115

*Fig. 7 — Folding Field Cultivator*

## Shanks

Spring-cushion shank mounting (Fig. 8) permits points to lift from 8 to 11 inches (203 to 279 mm), if they strike obstructions. Spring-cushioning also provides additional shank vibrating action for better soil pulverization, particularly in hard soil. Some spring-cushion mountings permit increasing shank rigidity for working extremely hard soil. Others permit adjusting shank pitch for normal, hard, and toughest soil conditions.

Spring-cushion shanks are standard equipment on some cultivators and optional on others, which may use fixed-clamp shanks (Fig. 9) for economy in soils with few obstructions. These shanks depend on flexing of the shank for protection from obstacles.

## Soil-Engaging Tools

A wide variety of soil-engaging tools is available for field cultivators. Double-point shovels, which can be reversed for longer wear, are most common in many areas. Sweeps from 4-1/4 to 12 inches (114 to 305 mm) wide are favored for other conditions. For economy, some machines with optional shank ends are forged into a chisel-type point (Fig. 10). When the original point wears down, reversible points may be attached for continued use.

JDPX6116

*Fig. 8 — Spring-Cushion Shanks Protect Equipment From Soil Obstructions*

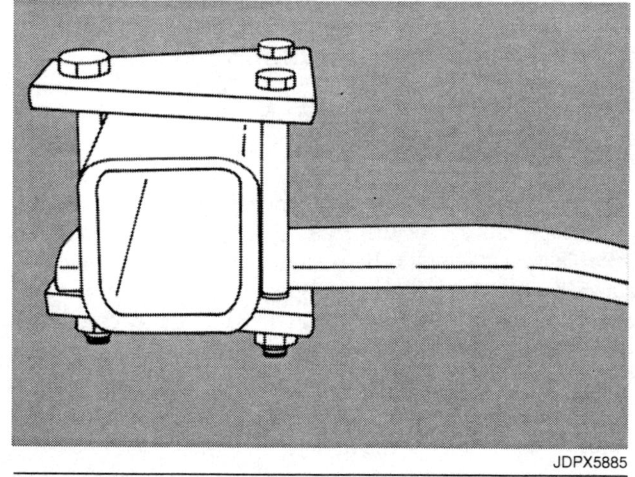
JDPX5885

*Fig. 9 — Fixed-Clamp Shanks for Economy and Obstruction-Free Soil*

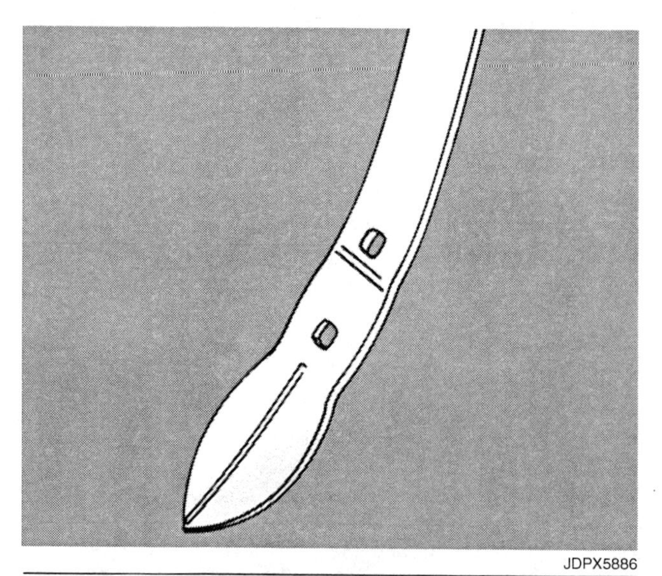

JDPX5886

*Fig. 10 — Shank End Forged Into Chisel Point*

Typical soil-engaging tools (Fig. 11) include:

- Narrow, reversible alfalfa point for renovation of alfalfa stands
- Double-point, reversible shovel (1-3/4 x 10-1/4-inch; 44 x 260 mm, shown), recommended for general tillage, hard ground, fall tillage or deep penetration
- 4-1/2 and 6-1/4-inch (114 and 165mm) sweeps for seedbed preparation and general tillage
- 9-, 10-, and 11-1/4-inch (229, 254, and 292mm) sweeps for killing persistent weeds and for fine seedbed preparation; sweeps should overlap for complete cutting of weed roots

## Principles of Field Cultivator Operation

Field-cultivator design, function, and performance are similar to those of chisel plows. (See Principles of Chisel Plow Operation, Chapter 6). Both can penetrate hard soil, break up large clods, and leave the surface open and broken to absorb moisture and resist erosion. The field cultivator is an excellent tool for seedbed preparation (Fig. 12) when equipped with sweeps, or smoothing-harrow attachment, and working light or previously tilled soil.

Varying shank spacing from the usual 6 inches (152 cm) to 9, 10, or even 12 inches (229, 254, or 305 mm) permits using field cultivators in heavy-trash or hard-soil conditions normally limited to chisel-plow operation. In extremely hard conditions, some shanks can be removed from each end of the cultivator to reduce draft load and still maintain desired operating depth.

If shank spacing is altered, or shanks are removed from the frame, the same number of equally spaced shanks must be retained on each side of the center line to provide uniform load and prevent side-draft.

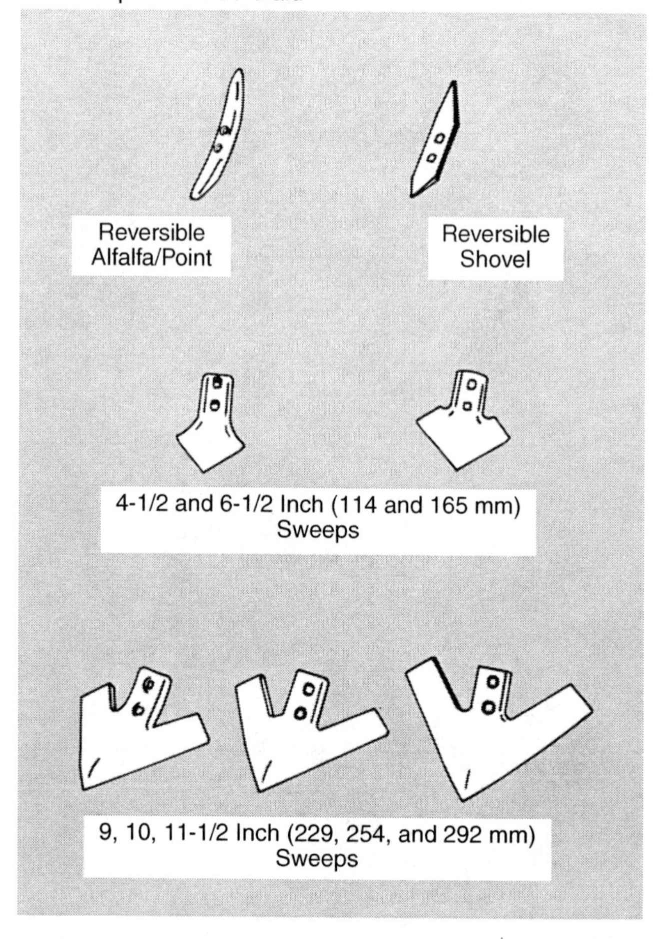


JDPX5887

*Fig. 11 — Typical Soil-Engaging Tools*

If all soil is evenly tilled, no particular field patterns need to be followed for field cultivators. Working on contours or across slopes helps reduce water runoff and erosion. Working at right angles to prevailing winds helps reduce soil blowing. Operating diagonally or at right angles to previous tillage operations provides better leveling and more uniform tillage.

JDPX6117

*Fig. 12 — Field Cultivators Prepare Seedbeds*

## Combination Implements

As reviewed in Chapter 6, several combination machines have been developed for the growing adoption of conservation tillage. These units appear very similar to mulch tillers, but are intended essentially for secondary tillage, seedbed preparation, and similar operations.

Several designs of combination implements are available. They typically include disk or coulter gangs in front, combined with a field cultivator. One design (Fig. 13) combines three tools into one implement: a disk gang that cuts and partially incorporates residue, a field cultivator equipped with wide sweeps that completes tillage, and a tine-toothed harrow that incorporates and smooths.

Another version combines a field cultivator equipped with sweeps and incorporator wheels. The latter lift previously buried residue back to the surface, incorporate chemicals through residue, and level the soil surface.

The amount of residue retained on the soil surface with these implements can be varied by choice of soil-engaging elements and depth of tillage. Factors affecting their operation and adjustments are similar to those outlined for field cultivators and disks. Instructions are provided in the operator's manual for each machine.

## Toothed Harrows

Implements used for secondary tillage differ somewhat by crop and geographic areas, but their wide range almost always includes some form of spike, spring, or tine-toothed implements.

These machines perform one or more of several functions: break soil crust, shatter clods, smooth and firm soil surface, close air pockets in the soil, kill weeds, loosen and aerate soil, conserve moisture, cultivate small plants, and renovate pastures.

## Tine-Tooth Harrow

Tine-tooth harrows, also called finger harrows, are relatives of spike-tooth harrows in appearance and performance under many soil conditions. The additional vibratory action of the tine teeth helps shatter clods, skip around obstructions, rip out small weeds, and smooth soil—and all of these are combined with light weight and low draft. Tine-tooth harrows are sometimes used for fast weeding and thinning of small sugar beet plants after emergence. Depth-control skids are available for many tine-tooth harrows to protect young plants when harrowed shortly after sprouting.

JDPX6163

*Fig. 13 — Combination of Disk, Field Cultivator, and Tine-Tooth Harrow*

JDPX6118

*Fig. 14 — Tine-Tooth Smoothing Attachments Improve Performance of Other Implements*

Tine-tooth sections are used as smoothing attachments for other implements such as mulch tillers, roller harrows, spring-tooth harrows, field cultivators, disks, and moldboard plows (Fig. 14). Very small tine-tooth sections also are available for use on corn planters to provide strip-tillage at planting time in previously worked soil. All these attachments provide additional soil pulverization and leveling for final seedbed preparation and normally eliminate at least one additional pass over the field, reducing soil compaction and saving time and fuel.

## Spike-Tooth Harrow

Spike-tooth harrows are also known as peg-tooth, drag section, or smoothing harrows (Fig. 15). They are used to smooth seedbeds, break soft clods, and kill small weeds as they emerge from the soil. Spike-tooth harrows are also used to break rain-crusted soil for fast emergence of seedlings.

A variation of the spike-tooth harrow, called a chain or flexible harrow (Fig. 16), is also used for seedbed preparation and other spike-tooth harrow functions—scratching up dead grass and loosening surface soil for seeding pasture or meadowland, scattering livestock droppings in pastures, and drying and smoothing dirt feedlots.

JDPX6210

*Fig. 15 — Final Seedbed Preparation With Smoothing Harrow*

JDPX5888

*Fig. 16 — Flexible or Chain-Type Harrow*

## S-Tine (Flexible) Cultivators

Breaking hard-crusted soil, preparing fine seedbeds, controlling weeds, renovating pastures, and cultivating row crops are only a few of the tasks performed by flexible or spring-tine cultivators.

The S or double-curved loop in the forged spring-steel tines provides a strong vibrating action (Fig. 17) that shatters soil, breaks clods, and pulls weeds out by the roots. These machines may also be used for early cultivation of row crops by relocating tines.

Teeth may do little more than scratch the soil surface at slow operating speeds. As speed is increased to 6 or 7 miles per hour (10 or 11 km/h), teeth draw into the soil and perform at their best. Even in hard, dry alfalfa stubble, soil may be broken to full depth and only small curved ridges left unbroken between tines.

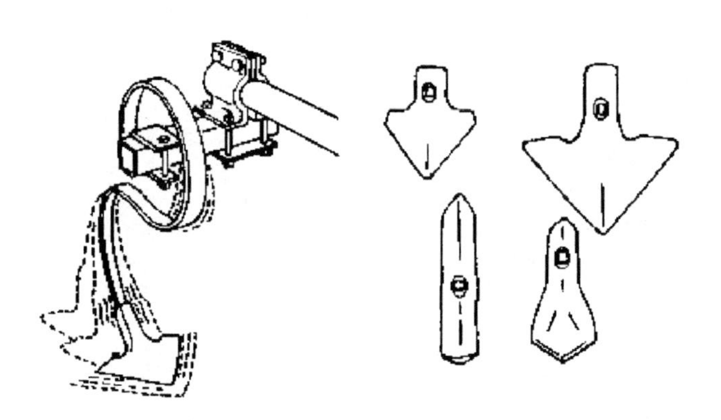

JDPX6119

*Fig. 17 — Vibrating S-Tine Shank Aid Swing of Tillable Sweeps*

High operating speed and strong tine action combine to make flexible-tine cultivators excellent for tearing out and destroying troublesome weeds and grasses.

In previously worked soil, flexible-tine cultivators break crusts and clods, kill weeds, close air pockets, and leave the ground ready for planting. Working depth is precisely controlled by adjustable gauge wheels or gauge shoes. Tractor load-and-depth or draft control need not be used with flexible-tine cultivators because draft requirement is quite low.

## Spring-Tooth Harrow

The spring-tooth harrow (Fig. 18) works 3 to 6 inches (76 to 152 mm) deep to loosen soil crust, dig, lift, and break clods that are not too hard, and pull many roots of quack grass and similar plants. Used immediately after plowing, it closes air pockets in the soil, breaks up clods, and levels the surface to make it ready to plant. In freshly plowed sod, operating in the direction of plowing with a shallow setting, the first time over, will help prevent dragging chunks of sod to the surface. The spring-tooth harrow is an excellent light-draft tool for summer-fallow and orchard and vineyard work.

Deeper penetration in crusted soil and more aggressive action on sprouting weeds makes the spring-tooth harrow much better suited for seedbed preparation than the spike-tooth harrow. However, large disk harrows and field cultivators have replaced spring-tooth harrows on many farms because of their greater ability to penetrate hard ground and kill tough weeds, particularly if the field cultivator has wide sweeps.

JDPX6165

*Fig. 18 — Spring-Tooth Harrow in Operation*

The spring-tooth harrow is better than the disk for stony ground, but plugs badly in heavy trash due to the limited clearance of most spring-tooth harrow sections.

## Field Conditioners

Field conditioners (Fig. 19) are essentially heavy-duty wheeled spring-tooth harrows—built wider and stronger. These lightweight machines are designed to make seedbeds, break soil crust, kill grass and weeds, control volunteer grain, incorporate chemicals and fertilizer, and work air pockets out of the soil. Working depth of most units is down to 6 inches (152 mm), depending on soil conditions. Vertical clearance is about 20 inches (508 mm).

Normal tooth-working interval on field conditioners is 6 inches (152 mm), which may be increased to 9 or even 12 inches (229 or 305 mm) in extremely heavy trash, or decreased to 4 inches (102 mm) with some units for seedbed work. Most machines have a standard two-bar frame with an optional third bar available to provide wider tooth spacing on each bar and still work soil completely.

JDPX6164

*Fig. 19 — Field Conditioner in Operation*

## Power Harrow

Utilizing tractor engine power through the PTO, rather than drive wheels, reduces or eliminates wheel slippage, reduces required tractor weight, and permits close control of the degree of tillage. The reciprocating action of the power harrow breaks clods and crusts without completely pulverizing soil when operated properly. Pointed tines work through the soil with no inversion and very little trash incorporation. Power harrows may overwork the soil unless properly used.

Moist lower layers of soil are not exposed, which means more moisture is retained for crop production. This also prevents exposure of large chunks of moist soil to rapid surface drying and formation of hard clods. By carefully controlling working depth, only drier surface soil is disturbed. Nor is there any slicing or sealing between layers of worked and unworked soil to restrict later water and root penetration.

## Roller Harrows and Packers

Roller harrows and packers are secondary-tillage tools which are related to various lighter types of field-finishing implements in that one of their purposes is to prepare seedbeds of proper tilth (Fig. 21).

Differences in construction and work between these tools and tooth-type harrows often are not great, but two generalizations apply: they usually are heavier than tooth harrows, and many of them compact the soil surface (even to a depth of several inches) instead of leaving it loose.

### Roller Harrow

Roller harrows are also known as cultipackers, cultimulchers, soil pulverizers, and corrugated rollers because of the resulting appearance of the soil surface. Other soil-firming implements that accomplish similar results, but usually under different conditions, are known as treaders, clodbusters, plow packers, and subsurface packers.

The primary purpose of these implements is to crush clods and firm the soil surface, which they do better than any other machines. Roller-type implements are useful for leveling and firming freshly plowed soil.

JDPX6211

*Fig. 20 — Roller Harrow*

JDPX5889

*Fig. 21 — Roller Packer Crushes Clods and Firms Soil Surface*

**Roller Packer**

Roller packers (Fig. 21) are also called land rollers, cultipackers, corrugated rollers, pulverizers, and packers. They also break soil crust, crush clods, firm the surface, and close air pockets near the surface for fast seed germination. With spring teeth raised, roller harrows may be used as roller packers to repack soil around roots of frost-heaved winter wheat, firm soil over newly planted crops, or break crusted soil for fast germination and growth if heavy rain immediately follows planting.

## Finishing Harrow (Do-All)

Another packer variation is the finishing harrow (Fig. 22), which has a pair of five-bladed, horizontal cutter heads to chop and crush clods and firm the soil. Cutter heads are followed by several rows of chain-suspended spike teeth to break additional clods and level the surface. Final finishing and smoothing is done by a hardwood leveling plank mounted across the rear of the machine. The finishing harrow (commonly called the do-all) is used primarily in the Mississippi Delta.

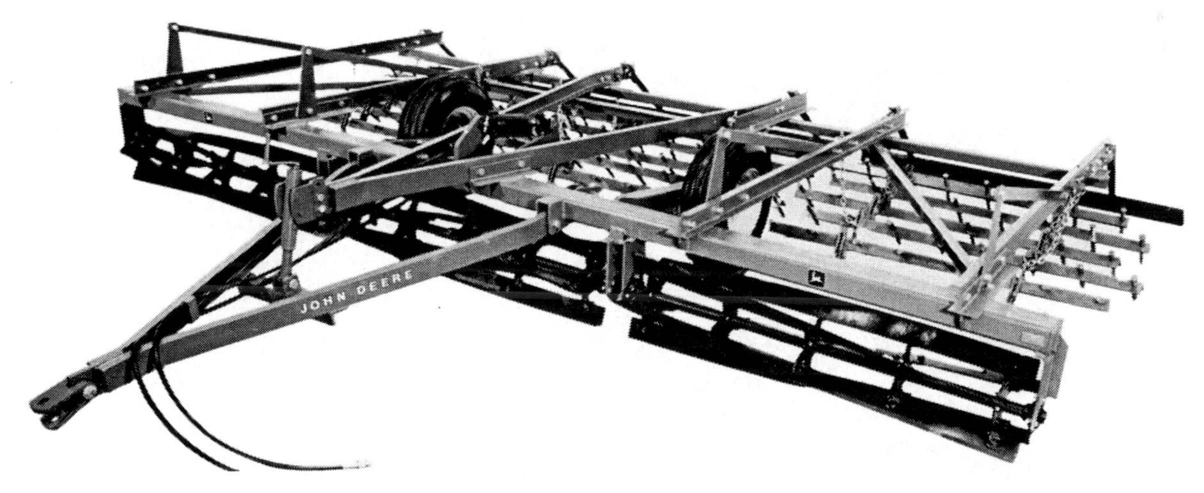


JDPX5890

*Fig. 22 — Finishing Harrow With Bladed Cutter Heads*

# Summary

Secondary tillage is used to prepare fields for planting, for summer fallowing, chemical incorporation, and weed control. The degree of tillage varies with the type of surface desired at planting.

Light to medium disks, field cultivators, combination implements, toothed and power harrows, and field conditioners are examples of secondary tillage equipment.

Light to medium disks are well suited for secondary tillage and are used more for this type of operation than for primary tillage.

Field cultivators have wider general utility than many other secondary tillage implements. Their uses range from seedbed preparation to weed and erosion control. They have a wide adaptability to tractor power, soil conditions, and tillage requirements.

Combination implements designed for secondary tillage typically include disk or coulter gangs combined with a field cultivator. Tines and incorporator wheels are available on some models. With choice of soil-engaging tools and depth of tillage, these implements can fit into virtually any secondary tillage system.

Some form of spike-, spring-, or tine-toothed harrow fits into almost all secondary tillage operations. They are used primarily for breaking clods or crusts, or controlling weeds, in final finishing of seedbeds.

Various configurations of roller harrows and packers are used to crush clods, firm the soil surface, and close air pockets to prepare seedbeds in clean-till systems for planting.

## Test Yourself

### Questions

1. List at least three purposes of secondary tillage.
2. (Fill in blanks.) Field cultivators are widely used for ________________, ________________, ______________, and ______________.
3. (Fill in blank.) Compared with other implements, field cultivators are most like ________________ in design and appearance.
4. Name two advantages of spring-cushion shanks for field cultivators.
5. (Fill in blank.) Combination implements for secondary tillage typically combine a _____________ or ________ with a ________.
6. Why are combination secondary tillage implements popular in conservation tillage systems?
7. How are tillage aggressiveness and residue retention controlled with combination implements?
8. Name three types of toothed harrows.
9. List three main uses of toothed harrows.
10. What is a field conditioner?
11. Roller packers are also called ________________, _____________, _____________ or ____________.
12. What are the tillage objectives of roller harrows and packers?
13. Name two basic differences between roller harrows and tooth-type harrows.

# Dryland Tillage

# 8

*Fig. 1 — Wide-Sweep Plow for Stubble-Mulch Tillage*

## Introduction

Western non-irrigated areas of the United States and Canada are characterized by low rainfall and frequent wind. Less residue from the previous crop is available than in more humid areas. Soil moisture in the seed zone at planting is often limited. Reduction of wind and water erosion and storage in the soil of all available moisture are primary objectives of tillage systems in the region. In the drier areas, summer fallowing is common. With this system of conservation farming, crops are grown in a field only in alternate years. Emphasis during the non-crop years is on controlling weeds, retaining residue cover, and storing soil moisture.

Stubble-mulch (wide sweep) plows, mulch treaders, rod weeders, disk tillers, and listers and bedders are examples of tillage tools that are used in the areas such as the western dryland region.

## Wide-Sweep (Stubble-Mulch) Plows

Wide-sweep or stubble-mulch plows cut off weeds at the roots and leave residue anchored to the surface with minimum soil disturbance. They are used for both primary and secondary tillage, either immediately after harvesting or just prior to planting. Sweep plows also are used to control weeds in summer fallow.

Wide-sweep plows have long V-shaped sweeps or straight blades that operate nearly at right angles to the direction of travel. Because of the wide shank spacing and relatively flat blade angle, these plows normally cover no more than 10 to 15 percent of the original residue with each pass (Fig. 1).

If the mulch is extremely heavy, it may be desirable to use a wide-sweep plow for early tillage and then mix some residue into the soil with a disk tiller or disk harrow before planting.

For maximum weed kill, plowing should be done on a hot day when soil is dry enough to crumble well. Under most conditions, speeds of 4 to 6 miles an hour (6.5 to 10 km/h) provide the best results and do the best job of loosening soil from weed roots.

In stubble, sweeps are normally operated just deeply enough to pass under the crown of the plant—about 3 to 6 inches (75 to 150 mm). If heavy accumulations of residue have been mixed previously with the soil, it may be necessary to operate deeply enough for blades to pass under it. Repeated plowings will work the residue back toward the surface.

JDPX6166

*Fig. 2 — Stubble-Treader Attachment for Wide-Sweep Plow*

The resulting mulch of stubble and dead weeds, plus the rough, shattered clods on the surface, reduce wind erosion and help catch and hold rainfall.

Improved weed control may be obtained by adding rod-weeder or stubble-treader attachments to wide-sweep plows (Fig. 2), or using them as separate implements. Additional weeds are killed and brought to the surface by these units, and residue is more firmly anchored to reduce blowing and water runoff.

Recently developed versions of the sweep plow consist of very wide flat sweeps on chisel plow shanks in sections 7 to 8 feet (2.1 to 2.4 m) wide for following uneven terrain (Fig. 3). These implements are used to replace the conventional stubble-mulch plow in primary tillage, to perform secondary tillage operations as shallow as 2 to 3 inches (50 to 75 mm), or to prepare seedbeds. They can be equipped with treaders that anchor residue firmly in place behind the sweep blades (Fig. 3) or with tine- or spike-tooth harrows for final soil conditioning before planting.

JDPX6167

*Fig. 3 — Sweep Plow on Chisel Plow Shanks*

### Types and Sizes

Wide-sweep-plow sizes range from one to nine sections with each section up to 8 feet (2.8 m) wide. Different combinations of blade size and number are available to match tractor power and field size. Fewer and wider sweeps may be used on some machines. The sweeps usually are staggered on the plow frame to permit smooth trash flow, and overlapped to ensure complete cutting of weed roots and to keep large weeds from slipping around the ends. Most common sweep sections are in the 5- to 7-foot (1.5 to 2.1 m) range.

Wide-sweep-plow frames are much like chisel-plow frames, but somewhat stronger, with extra reinforcement at stress points and for shank attachment.

Small wide-sweep plows may have one or two sweeps, and transport width is essentially the same as cutting width. Larger units have wings with one or two sweeps each. These wings are folded hydraulically for transport, and are attached to a center section.

Some six-section models have two wings on each side, with the outer sections folded flat across the top for transport (Fig. 4). This reduces transport height as well as width.

One or more remote hydraulic cylinders control working depth, raise and lower the machine, and fold wings for transport. Optional flotation tires on some machines provide increased stability in soft or loose soils.

Rolling coulters, usually equipped with rippled blades for more positive cutting and turning, slice through residue directly in front of each shank to prevent buildup and plugging. This means that more field time is spent plowing —less time spent unplugging.

*Fig. 4 — Six-Section Plow Folded for Transport*

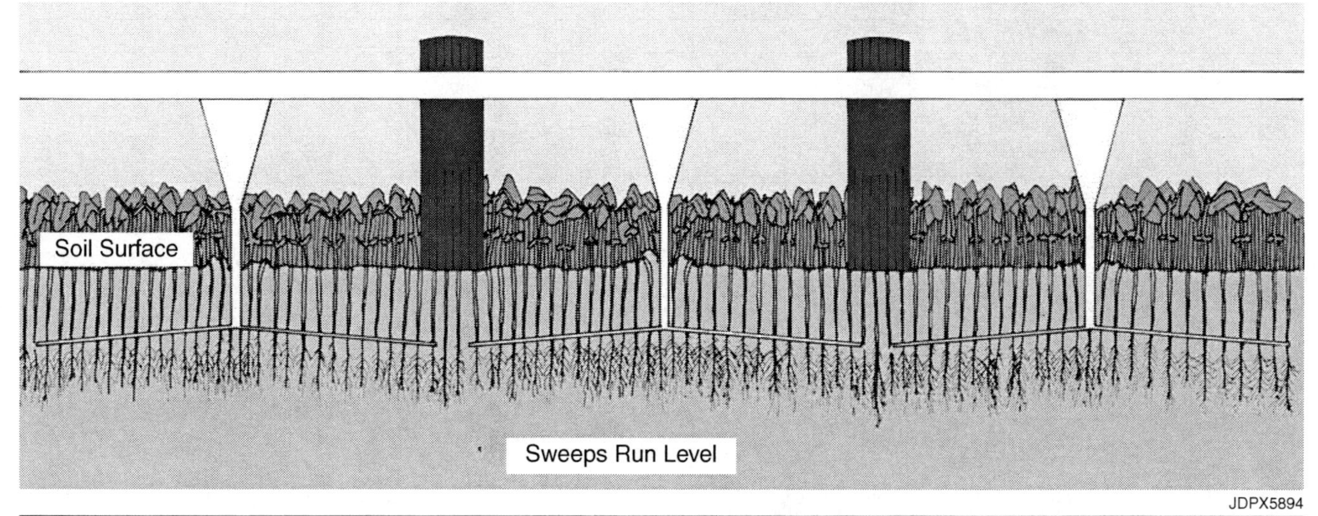


*Fig. 5 — Blades Run Level at Equal Depth*

## Field Operation

Operating characteristics of wide-sweep plows and chisel plows are similar, and both machines are handled in much the same manner. Sweep-plow performance is best and draft lowest when all blades are running level from side to side and at the same depth (Fig. 5). Sweep points should either be set level with the blade ends, or as much as a half-inch (13 mm) lower to aid penetration in hard soils.

Shims or other means of adjustment between the main frame and shanks permit leveling sweeps with the frame. Hitch adjustments are usually provided for fore-and-aft leveling. The hitch setting must be readjusted when changing operating depth or switching to a tractor with different height drawbar. Adjustable links between the center and wing sections permit lateral leveling.

Adjust the rolling coulters to work directly ahead of the sweep shanks and just deep enough to cut through trash and keep turning.

Sweep blades must be sharp to be able to penetrate hard soil. If penetration remains a problem after sweeps have been sharpened and leveled, it may be necessary to add additional weight to the frame.

**IMPORTANT: Always fasten weights securely to the frame to prevent shifting, bouncing, and possible equipment damage. NEVER raise wings unless weight is removed or firmly attached to the plow frame.**

## Rod Weeders

Rod weeders (Fig. 6) are widely used in western wheat-growing areas of the United States and Canada, primarily for weed control in summer fallow and prior to seeding. Weeding can be done shortly after seeding, before germination and root development, by setting the rod far enough above planting depth to root out weeds without disturbing seed.

The rotating weeder rod is operated from just under the soil to several inches deep to pull weeds out by the roots and work them to the top, along with coarser soil particles, to provide a surface mulch.

Rod weeders normally cover only about 10 percent of the crop residue each pass, which makes them excellent for stubble-mulching where maximum residue should be left exposed to catch and hold moisture and resist runoff and soil blowing.

In field operation, rod weeders are typically worked across slopes or at right angles to prevailing winds to reduce water runoff and soil blowing.

*Fig. 6 — Rod Weeder*

## Principles of Rod-Weeder Operation

Weeder rods may be round or square, and more than one rod size is available for each machine—a smaller one for economy and a heavier one for increased strength for severe conditions. Typical options are 7/8- and 1-inch-square (22.2 and 25.4 mm) high-carbon steel rods.

The front side of the rod moves upward as it revolves, to pull up and get rid of roots while leaving most of the surface residue intact (Fig. 7). This requires a reversing drive in the boot. As the machine moves forward, the drive axle rotates (left), turning the double sprocket at the top of the boot. This powers the rod drive chain, which runs down through the boot to rotate the rod in the proper direction. Square rods simply slide through square holes in the rod sprocket and are clamped in white-iron bearings on each pendant. Round rods must be drilled or keyed to hold the sprockets in place.

*Fig. 7 — Rod Revolves to Pull Up Roots*

### Disk Tillers

Disk tillers fall between the disk harrow and moldboard and disk plows in function (Fig. 8). They consist of spherical blades, which normally throw the soil to the right, mounted on a common axial shaft (or gang shafts for flexible types).

Disk tillers also are known as wheatland disk plows, one-ways, one-way disks, diskers, seeding tillers, vertical disk plows, and by other names. They were developed in the Great Plains area about 1927.

The normal function of a disk tiller is to cut and mix soil and plant residues with a minimum of soil pulverization. A considerable amount of residue is normally left protruding from the tilled soil to control wind and water erosion. Depth may be varied from 2 to 8 inches (50 to 203 mm), depending upon blade size and spacing.

Disk tillers are normally used for primary tillage, and in subsequent operations for summer fallowing. Their weed control is excellent. By adding seeding and fertilizing attachments, they may be used to prepare a seedbed, seed, and fertilize in a single, efficient operation.

### Listers and Bedders

Lister, bedder, lister-planter, lister-bedder, disk bedder, middlebuster, and middlebreaker are terms used in different areas to describe the same equipment. The local name usually pertains more to the operation being performed than to the implement, and much confusion in names has come about from different applications of similar tools. Reshaping of existing beds is sometimes called hipping.

JDPX6098

*Fig. 8 — Disk Tiller in Operation*

Bedders (Fig. 9) are normally classified as primary-tillage tools that open furrows or make beds. Lister-planters (Fig. 10) are primary-tillage tools that not only open furrows but plant simultaneously in the bottoms of the furrows.

JDPX6212

*Fig. 9 — Bedders Shape Ridges but Do Not Have Planting Attachments*

JDPX6213

*Fig. 10 — Bedder Planters Shape or Reshape Beds at Planting*

These implements may have solid moldboard-type bottoms or small disk gangs (Fig. 11). Choice depends on soil and moisture conditions and operator preference. Disks may work better in soils where lister bottoms do not scour well or if there are many soil obstructions. Lister bottoms develop suction to aid penetration.

Disks have little or no suction and often are used after previous primary tillage has loosened the soil. Lister shares may need to be sharpened or replaced more often than the disks.

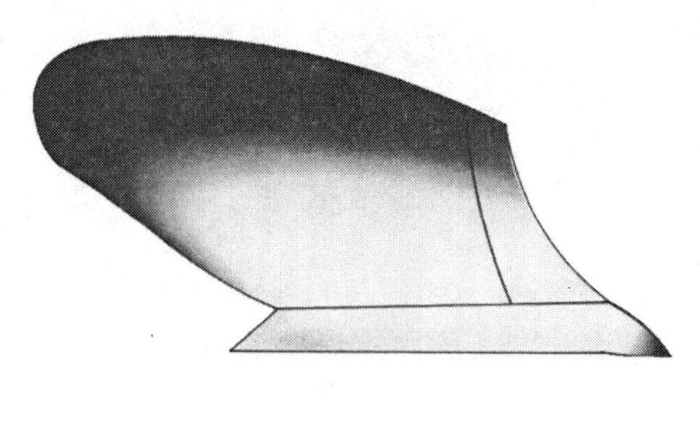

JDPX5871, 6214

*Fig. 11 — Listers and Bedders May Have Disks or Moldboards*

Lister planting is usually done in dryland area to place seed in moist soil for fast germination and quick growth. It is used in parts of the South to plant double-crop soybeans or sorghum following small-grain harvest.

Such planting reduces the time between grain harvest and planting to provide maximum growing season for the second crop and conserve available soil moisture by reducing tillage to a minimum.

### Profile Planting

A variation of lister planting, sometimes called profile planting, puts the crop row on a small ridge or raised bed in the bottom of the lister furrow (Fig. 12). This places seed in moist soil, protects young plants from wind and blowing soil, and reduces the chances of water standing on the row or washing over the seed or seedlings during a heavy rain.

In some areas beds are shaped soon after fall harvest and disposal of crop residue. If the soil was previously listed or bedded, the old ridges or beds are usually split and new ones formed over the old furrows. Before planting, the new ridges may be reshaped, or split again (Fig. 10), and re-formed in the same area as ones of the previous year. Fertilizer may be applied under the row area during the listing operation.

If the land has been listed for some time and soil has eroded into the furrows, ridges may be reshaped and furrows cleaned with the lister. Crops to be planted, soil conditions, and local custom usually dictate the practices and number of operations performed in preparing land for planting.

## Planting Methods

Beds may be planted in various ways. A unit planter with a large, specially shaped sweep (Fig. 6) may be used to cut away dry soil on top of the bed and plant a single row of corn, sorghum, cotton, or other crop on each bed. In drier areas the furrows can be used for later irrigation.

### Ridge or Bed Planting

Planting on top of beds is common in high-rainfall areas where surface drainage is a problem. Soil in beds dries and warms faster than in furrows or on flat land, permitting faster seed germination and better early crop growth. This prevents water from standing on rows after heavy rain and possibly killing small plants. Depending on the bed-shaping and cultivation methods used, weed control may be more of a problem. Higher herbicide application rates may be required.

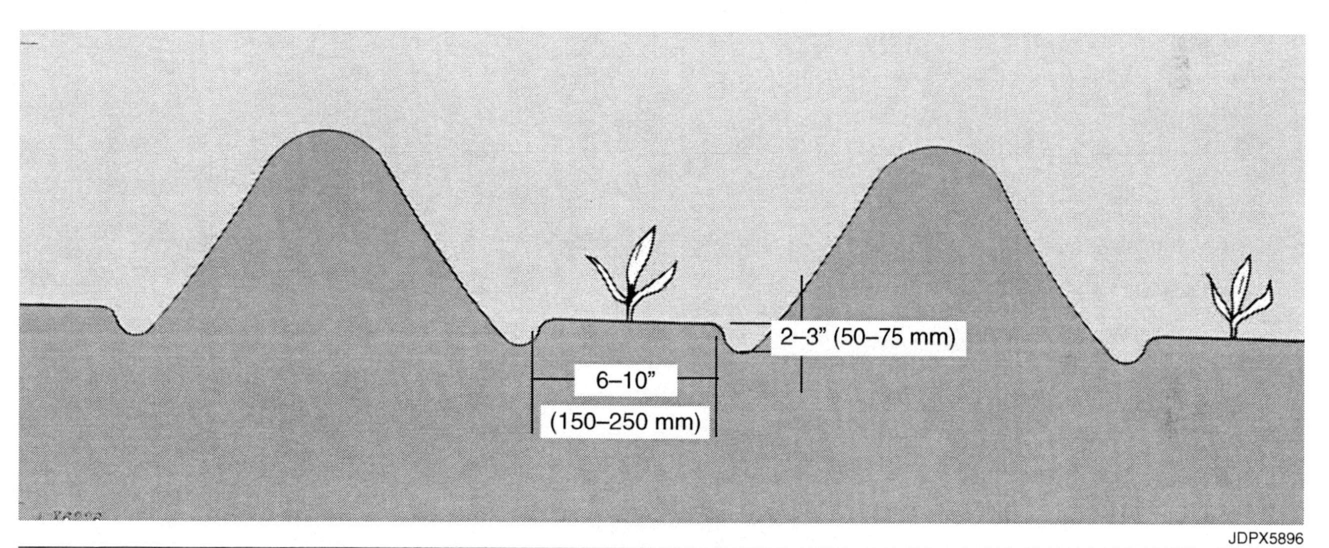


*Fig. 12 — Profile Planting Protects Seed and Small Plants From Heavy Rain and Wind*

## Summary

Wide-sweep (stubble-mulch) plows, mulch treaders, rod weeders, disk tillers, and listers and bedders are used in areas where low moisture and wind erosion are problems.

The V-shaped blades of wide-sweep plows kill weeds with minimum disturbance of residue and minimum loss of moisture. They also loosen the upper 3 to 6 inches (76 to 152 mm) of soil for better rainfall infiltration. These plows can also be equipped with treaders that anchor residue firmly in place or with tine- or spike-tooth harrows for final soil conditioning before planting.

Rod weeders have round or square rods that turn backward under the surface to pull weeds out by the roots and also to work weeds and coarse soil particles to the surface as an erosion-resistant mulch.

Disk tiller blades mix residue into the layer. They should not be used where they produce a powdery surface.

Listers and bedders can be used to combine tillage and planting into one operation in southern states and drier western areas by forming ridges or beds, then planting in the furrow between beds, or on top of the beds.

## Test Yourself

### Questions

1. What is the typical operating depth of wide-sweep plows?

2. (Select one.) HOT or COOL days are best for using wide-sweep plows.

3. (Fill in blanks.) Mulch treaders add to wide-sweep plow effectiveness by ___________ and ________________.

4. (T/F) The normal function of a disk tiller is cutting and mixing of soil, with minimum pulverization.

5. (Fill in blanks.) ____________ are normally classified as primary-tillage tools that open furrows. ___________ are primary-tillage tools that not only open furrows, but plant simultaneously in the bottoms of the furrows.

6. What is profile planting? What are its advantages?

7. Where is planting on top of beds a common practice?

8. (T/F) The bedder is normally classified as a secondary-tillage tool.

9. In what conditions are disk bedders superior to bedders with moldboard-type bottoms?

10. Which way does the weeder rod rotate?

11. When and where are rod weeders usually used?

# Tillage at Seeding

# 9

## Introduction

Conservation tillage systems are designed to retain a considerable amount of surface residue from the previous crop. Planters, drills, and seeders must open furrows and place and cover seeds while leaving the soil surface as undisturbed as possible. Some tillage, therefore, is required at seeding. With no-till systems, no prior tillage is done except for a narrow strip in front of the planting machine opener.

Tillage at planting enables the farmer to reduce the number of trips over the field. Also, shortage of farm labor and larger farm sizes have increased the popularity of combining tillage and planting in both full-tillage and conservation farming systems.

Heavy-duty row-crop planters, grain drills, and seeders have been developed specifically for satisfactory planting in high residue or no-till conditions. Tillage attachments are also available for conventional planting equipment that provide for planting in many tilled and untilled conditions.

## Row Crop Planters

Modern row-crop planters are designed for performance in seedbeds that range from clean-tilled to heavy-residue, conservation-type conditions (Fig. 1). Hinged and folding frame designs permit consistent planting over terraces and contours and also narrow transport width of wide planters.

JDPX5897, 5898

*Fig. 1 — Row-Crop Planters Operate Well in Both Clean-Tilled and Conservation Seedbeds*

A till planter (Fig. 2) is available for use in tillage systems where the soil between rows is not tilled. The system often includes chopping stalks in the fall, then tilling in the row and planting as one operation in the spring. Fall stalk chopping is sometimes omitted and stalks are sliced with a coulter at planting. Best results are usually obtained if planting is done on beds prepared during cultivation of the previous crop.

JDPX5899

*Fig. 2 — Till Planter*

Several types of tillage attachments are available to improve planter performance in tilled or untilled seedbeds:

- Disk furrower attachment

A common practice in some areas is to run a planter behind a cultivator on fall-plowed land. The disk furrower attachment (Fig. 3) pushes clods from in front of the furrow opener to provide more uniformity in planting depth.

JDPX6215

*Fig. 3 — Disk Furrower Attachment*

- Fig. 4 shows a Tru-Vee opener that slices a narrow, V-shaped furrow creating a prime environment for seed. The two disk blades easily carve through hard soil and heavy residue.

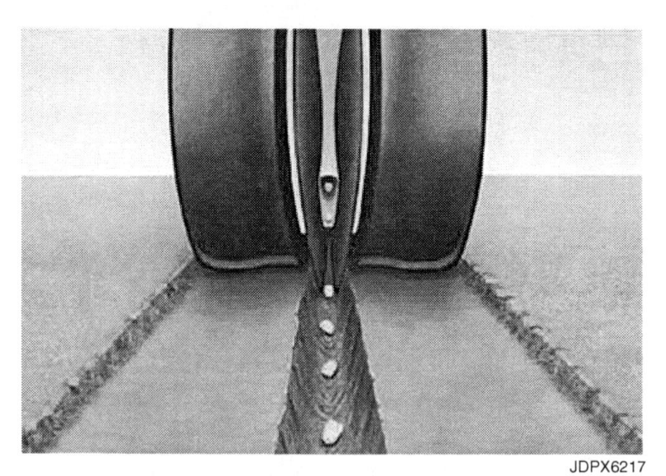
JDPX6217

*Fig. 4 — Tru-Vee Opener Attachment*

- Tine-tooth tillage attachments (Fig. 5) prepare only the seedbed or row area.

JDPX6168

*Fig. 5 — Tine-Tooth Attachment*

- Fluted, ripple, and serated coulters are no-till planter attachments that help prepare the seedbed without prior tillage.

The fluted coulter (Fig. 6) prepares a seedbed about 3 inches (7.5 cm) wide and 2 to 3 inches (5 to 8 cm) deep.

The ripple coulter may be attached to the planter as shown in Fig. 5. This coulter prepares a seedbed about 0.75 to 1.5 inches (1.9 to 3.8 cm) in width.

JDPX6216

Fig. 6 — Fluted Coulter for No-Till Planter

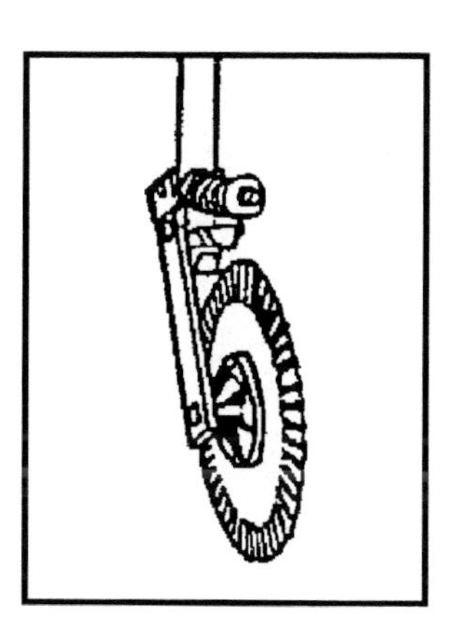
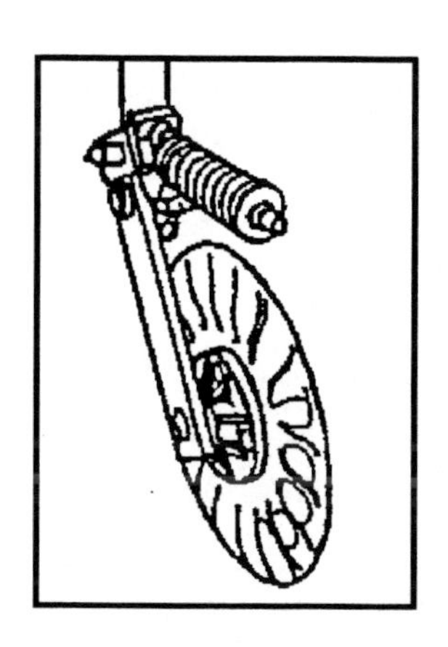

JDPX6218

Fig. 7 — Ripple Coulter for No-Till Planter

**Row Cleaner Attachment**

Surface residue on front of the planter can hairpin at the disk opener and cause poor soil penetration. The row cleaner attachment (Fig. 8) is designed to sweep away the residue for improved opener performance.

JDPX6219

Fig. 8 — Row Cleaner Attachment

## Air Seeders

Air seeders were developed in Germany several years ago for seeding small grain. The concept has become popular in areas where grain is produced on large flat areas of land, such as the wheat areas of Australia and the Great Plains of the United States and Canada. More recently, it has been considered as a potential soybean seeding alternative.

JDPX6122

*Fig. 9 — Field Cultivator Air Seeder*

The seeders consist of two separate implements: a seed tank and a tillage tool, such as a chisel plow (Fig. 9), field cultivator, or sweep blades, operating together to form one seeding tool. This combination allows the field operations of tilling, seeding, and fertilizing to be completed in one field pass.

The air seeder has the basic components of any seeding machine, including furrow opener, and seed-metering and seed-placing devices (Fig. 10).

JDPX6123

*Fig. 10 — Air Seeder Components*

Air seeders feature an air delivery system for uniform seed distribution from the metering mechanism to the seedbed. The air stream is a very efficient method of transferring the seed fertilizer from a central seed tank to a number of remote locations on the tillage shanks.

The central seed tank provides greater capacity, resulting in fewer stops to refill the tank. It can be towed behind or in front of the seeding tool.

Different types of furrow openers are available, depending on the condition of the soil and the amount of tillage desired. Sweeps are designed to give level cutting action. Chisels may be equipped with a fertilizer boot and used for fertilizer banding.

## No-Till Drills

While they perform equally well in fully tilled seedbeds, no-till drills were designed for planting in no-till, minimum-till, or other conservation tillage systems where sod, large amounts of surface residue (such as corn stubble), or very firm soil would limit conventional types of drills (Fig. 11).

For seeding in difficult conditions, no-till drills are equipped with heavy-duty frames and single disk openers with adjustable down pressure for penetrating hard soil. They also feature high clearance for operating in heavy residue (Fig. 12).

Gauge wheels mounted beside each disk opener provide for uniform seed placement in uneven fields and high-residue situations. Semi-pneumatic press wheels firm the seed in the bottom of the furrow for close seed-to-soil contact and good seed germination.

Grain/fertilizer models of no-till drills also permit fertilizer application in the furrow with seed or in a separate furrow away from the seed. These drills can also be used for placing fertilizer only.

JDPX5900

*Fig. 11 — No-Till Drill*

No-till drills are often considered as small grain equipment only. While they are popular in small grain areas, they have significantly influenced soybean seeding methods in more humid areas. Soybeans are seeded with these drills under conditions that would formerly have required near full tillage. Some farmers seed soybeans without prior tillage; others till in the fall using equipment such as a mulch tiller, then seed in the spring with a no-till drill.

JDPX5901

*Fig. 12 — No-Till Drills Are Designed for Difficult Seeding Conditions*

## Press-Wheel Drills

The press-wheel drill is also useful for planting in almost any seeding condition from clean till to no-till (Fig. 13).

A press-wheel drill has press-wheel gangs mounted on the rear of the drill. The press wheels firm the soil over the seed, drive the seed metering mechanism, and support the rear of the drill.

*Fig. 13 — Press-Wheel Drill*

Most models of press-wheel drills can be equipped with double disk openers for clean-till or light-residue conditions. More difficult conditions require hoe openers.

Several types of hoe openers are available for different seedbed conditions (Fig. 14). The double-point shovel is used for shallow seeding with minimal disturbance of soil. Spear-point shovels provide penetration in dry seeding conditions and create a ridge. The wrap-around single-point shovel creates a larger ridge than the double-point shovel. The 4-inch (102 mm) shovel provides tillage and ridging action for controlling wind erosion.

Several models of press-wheel drills permit application of fertilizer while seeding or metering grass and grain seeds in the same pass.

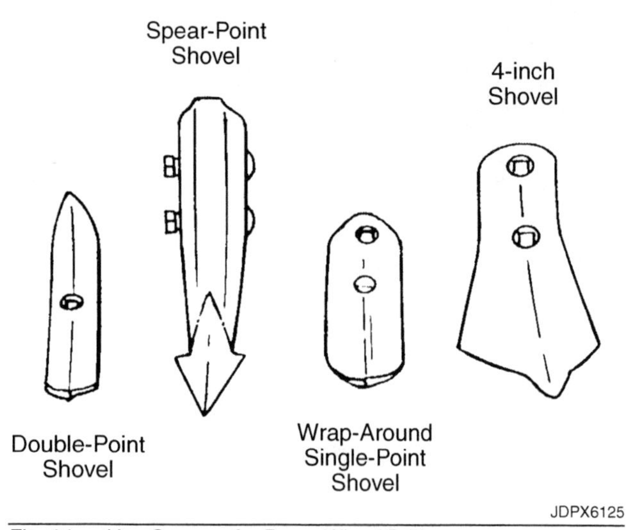


*Fig. 14 — Hoe Openers for Press-Wheel Drills*

### Tilling Seeder

The tilling seeder (Fig. 15) is an end wheel type drill with power-driven cutter wheels that prepare a seedbed for each seed drop on the drill. The cutter wheels run in front of the seed drops on steel seed boots.

*Fig. 15 — Tilling Seeder*

The tilling seeder can be used to plant new pastures, renovate old pastures, and seed roadsides, parks, and golf courses. Seeding with the tilling seeder eliminates tearing up existing sod and cuts trips over the field to interseed, as with conventional pasture renovation methods.

## Summary

Conservation tillage systems retain large amounts of surface residue from the previous crop. Planters, drills, and seeders must open furrows, then place and cover seeds in these conditions while leaving the soil as undisturbed as possible. Therefore, a minimal amount of tillage is required at seeding, even with no-till systems.

Heavy-duty seeding equipment is available for these difficult seeding conditions. Equipment to provide tillage at seeding includes the row-crop planter, till planter, air seeder, no-till drill, press-wheel drill, and tilling seeder.

Most of these tools perform well in fully tilled seedbeds, but were designed to seed in high-residue, no-till or reduced tillage, hard soil conditions. Several types of coulters and furrow openers are available to meet specific seeding requirements.

## Test Yourself

### Questions

1. List four types of tillage attachments for row-crop planters.

2. What is the difference between a ripple and fluted coulter?

3. List two functions of press wheels on press-wheel drills.

4. Name three applications for the tilling seeder.

5. (Fill in blanks.) Air seeders consist of two specific implements: ________ and a ____________ such as __________, __________ ,or _____________.

6. (T/F) Fertilizer may be applied with a no-till drill in the seed furrow or in a separate furrow.

7. How does the method of seed delivery with an air seeder differ from that of a grain drill?

8. (T/F) Seeding equipment that provides tillage at planting can be used successfully only in conservation tillage systems.

9. (Fill in blanks.) The no-till drill is used primarily to seed _________________ and ___________________.

10. (T/F) The seed tank in an air seeder system must be towed behind the seeding tool.

# Post-Emergence Tillage

# 10

## Introduction

Weed control, preparation of land for irrigation, and breaking soil surface crust are primary purposes of post-emergence tillage.

Sweeps, shovels, and rotary-cultivator or rotary hoe wheels slice off weeds, pull small weeds, and shatter the soil surface to expose additional weed roots to die in the sun. When using small sweeps and shovels, some well-rooted weeds are apt to slip around the tool without being cut. This can be minimized by overlapping large sweeps or weeding knives, and must be considered in arranging tools on the cultivator.

Weeds usually germinate in the upper inch or two (25 to 50 mm) of soil, so shallow cultivation will normally control those weeds and reduce germination of additional seeds that are buried deeper. High concentration of soybean feeder roots have been found in the top 2 to 4 inches (50 to 100 mm) of soil. Corn roots will quickly spread across the entire inter-row area in uncompacted soil. Cutting the plant's roots with deep cultivation reduces the crop's ability to gather moisture and nutrients from the soil and can seriously reduce yields.

JDPX6127

*Fig. 1 — Rear-Mounted Row-Crop Cultivator*

Soils that tend to crust seem to benefit most from cultivation. Stirring the soil surface breaks up the crust, improves aeration and moisture absorption, and kills weeds growing between rows. Cultivation also helps control weeds that may be resistant to most herbicides, and may discourage germination of some late-starting weeds, such as fall panicum.

## Types and Sizes of Post-Emergence Tillage Equipment

Equipment typically used for post-emergence tillage includes front-, mid-, and rear-mounted row-crop cultivators (Fig. 1 and Fig. 2) rotary cultivators, and rotary hoes.

JDPX6231

*Fig. 2 — Mid-Mounted Row-Crop Cultivator*

## Row-Crop Cultivators

Size of row-crop cultivators is counted by the number of rows covered and ranges from four-row units up to twelve 40-inch (102 cm) or sixteen 30-inch (76 cm) rows. The size selected depends on area to be tilled, tractor power, tractor lift capacity, and time available for tillage without interruption by weather or other farm operations. The most common type of front- and rear-mounted cultivators has either one or two independent gangs or rigs attached to the toolbar with parallel linkage.

Rear-mounted cultivators with up to eight 40-inch (102 cm) rows can be transported on the tractor 3-point hitch without change. Wider units may be equipped with an end-transport attachment and reduced to less than 8 feet (2.4 m) in width for transport. Others are mounted on folding toolbars so end sections may be folded for transport (Fig. 3).

Stabilizing coulters on rear-mounted cultivators provide lateral stability and help average out minor steering corrections as the cultivator follows the tractor.

Some models of rear-mounted cultivators are designed for use in reduced tillage systems or other heavy-residue conditions (Fig. 5). These cultivators have heavier construction and high clearance. Rolling shields maneuver through stalks and other residue to protect small crop plants as the cultivator moves over the field.

Heavy-duty cultivators with chisel plow shanks have the capability of working in high-residue or other difficult conditions such as no-till fields (Fig. 4). When equipped with ridging wings, they can build ridges while cultivating in ridge-till systems.

## Front-Mounted Cultivators

Cultivators of up to 12 rows may be front mounted on many tractors. In contrast to rear-mounted cultivators, front-mounted units are rigidly attached to the tractor laterally and shift directly with changes in steering direction.

More continuous steering attention is required with these cultivators because they do not have the lateral flexibility of rear-mounted equipment to average out steering corrections or deviations. Visibility of the row units is better with front-mounted cultivators, but is often over-stressed in relating one type to the other, particularly on wider cultivators.

JDPX6128

*Fig. 3 — Wider Cultivators Can Be Folded for Transport*

JDPX6169

*Fig. 4 — Cultivator for No-Till or Ridge-Till Conditions*

JDPX6170

*Fig. 5 — Cultivator for Reduced Tillage Systems*

## Rotary Cultivator

Rotary cultivators (Fig. 6) are versatile implements made of a series of spider gangs that look much like rotary-hoe wheels. The curved rotary teeth slice and twist as they pass through the soil, uprooting small weeds, cutting large ones, and breaking soil crust. Contrary to rotary-hoe operation, rotary cultivator wheels turn backward and have curved slicing teeth. Teeth are designed for either right- or left-hand cutting. Gangs are individually angled for moving soil toward or away from rows, working on the sides and tops of beds, or in combination with sweeps and shovels. Rotary gangs are also used for incorporation of chemicals at planting time. By arranging gangs for complete soil coverage, they can be used to incorporate broadcast chemicals. Gangs can be set close to rows for clean cultivation of small crops, and later turned to throw soil into the row to cover late weeds.

Recommended operating speeds, usually 5 to 8 mph (8 to 13 km/h), will vary with soil conditions, crop size, and operator experience.

JDPX6129

*Fig. 6 — Rotary Cultivators Can Work up to 8 Miles per Hour (13 km/h)*

## Rotary Hoe

About the fastest, most economical means of cultivation is rotary hoeing (Fig. 7), which can be done prior to crop emergence, or after plants are at least 2 inches (50 mm) tall, to break up soil crust. One or two passes with the rotary hoe when the crop is small often eliminates later cultivations with other equipment.

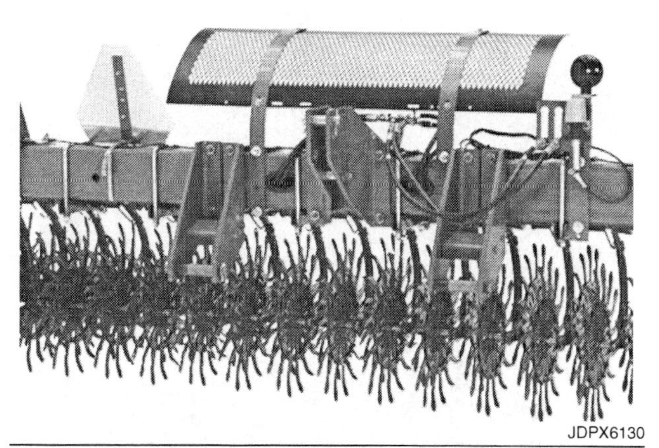

JDPX6130

*Fig. 7 — Rotary Hoe Provides Fast Early Cultivation*

The points of rotary hoe wheels enter the soil almost vertically and emerge pointing rearward with a lifting action which shatters crusts, pulls or covers small weeds, and destroys weeds that have not emerged. The rotary hoe may be used only once or whenever a new crop of weeds emerges. It usually will not destroy weeds or grasses after roots are established.

Optimum performance is at operating speeds of 7 to 12 miles per hour (11 to 19 km/h). The recommended top and minimum speed varies by manufacturer. High-speed action pulls many weeds free of the soil and leaves them exposed to die. Extremely high speeds may tear leaves from crop plants and cause some uprooting, but the benefits from rotary hoeing more than offset any crop mortality.

Rotary hoes are available as integral units in sizes from four to twelve 40-inch (102 cm) rows or six to sixteen 30-inch (76 cm) rows. Most models have hoe wheels attached in pairs to spring-loaded shanks. One wheel works slightly ahead of and 3-1/2 inches (89 mm) to one side of its mate. The arm connecting the two wheels permits them to rise and fall individually over uneven ground or field obstructions, or to move together on the shank.

Spring-loaded shanks are mounted on a common 15 to 30-foot (4.5 to 9 m) integral toolbar (Fig. 7) to provide full coverage for up to twelve 30-inch (762 mm) rows. When the hoe is raised to transport position, spring pressure on the shanks is relieved and shanks drop almost straight below the bar.

An endways transport attachment permits quick changeover from working position to transport and reduces width to less than 8 feet (2.4 m). Optional frame-stabilizer wheels are available for outer ends of the toolbar to maintain proper height in operation. These same wheels can be used for the transport attachment.

JDPX6131

*Fig. 8 — Spring-Top Shanks Automatically Reset After Passing Over Obstructions*

## Cultivator Attachments

Most front- and rear-mounted cultivators use the same soil-engaging attachments, with only slight variations for some applications.

Several shank types are available, including spring-trip (Fig. 8), which trip and automatically reset when an obstacle is encountered, and quick-adjustable and friction-trip shanks for soils with fewer obstructions.

Row shields are required when cultivating small crops to prevent covering plants with soil. Some shields are attached directly to the rig (Fig. 9) on parallel linkage to permit vertical flexing over obstructions and maintain level operation.

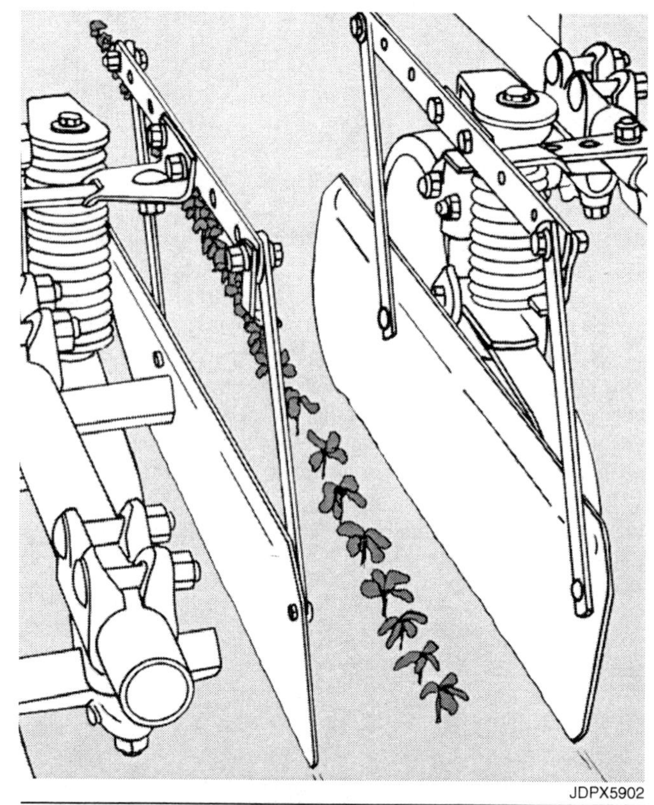

JDPX5902

*Fig. 9 — Rig-Attached Plant Shields Are Adjustable for Row Spacing and Crop Conditions*

Rolling shields (Fig. 10) are particularly helpful when cultivating small plants in heavy trash, because they roll over instead of dragging trash. They're ideal for high-speed cultivation.

Hooded shields (Fig. 11) are fully adjustable and provide maximum possible plant protection. Rotary cultivator shields protect bushy crops, such as soybeans, and may be moved up or down to control flow of soil to the plants.

Spray shields (Fig. 12) include two nozzles for directed herbicide application in the row area while middles are being cultivated. Spray is held within a controlled area.

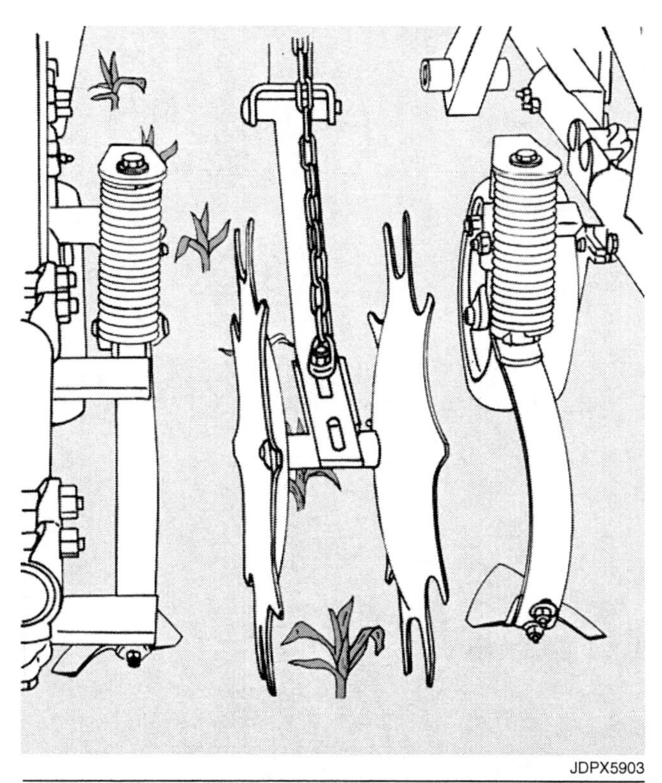


*Fig. 10 — Rolling Shields Work Well in Trash and for High-Speed Cultivation*

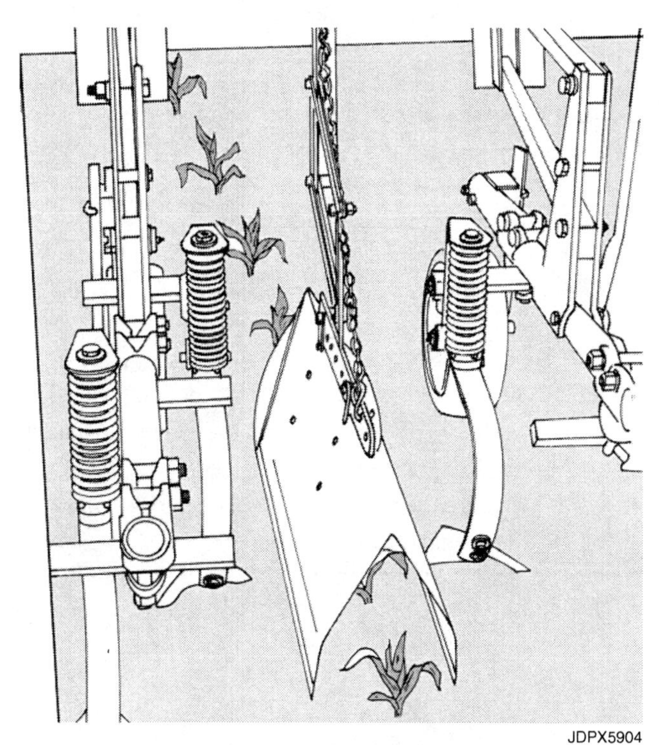


*Fig. 11 — Hooded Shields Provide Maximum Crop Protection*

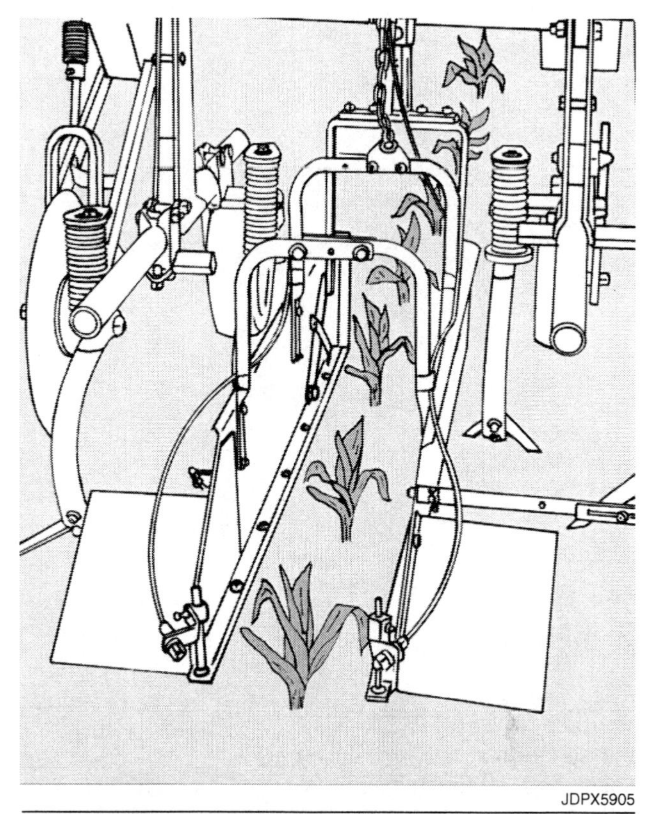


*Fig. 12 — Spray Shields Permit Herbicide Application While Cultivating*

Cultivator soil-engaging tools range from 24-inch (610 mm) wide sweeps to 16-inch (406 mm) diameter disk hillers, their use depending on crop, soil, and moisture conditions. Some tools are available with hard-facing for longer wear in abrasive soils. Typical cultivator sweeps (Fig. 13) include:

- Universal sweeps low crown, narrow shank, low wing angle, and quick scouring for high-speed cultivation
- Peanut sweeps for high-speed cultivation with minimum soil movement
- Mixed-land sweeps—high, broad crown, high wing angle, and broad shank to throw soil and leave a mulched surface
- Blackland sweeps—high crown and wing angle for difficult scouring conditions
- General-purpose sweeps—low-pitched wings for good soil flow

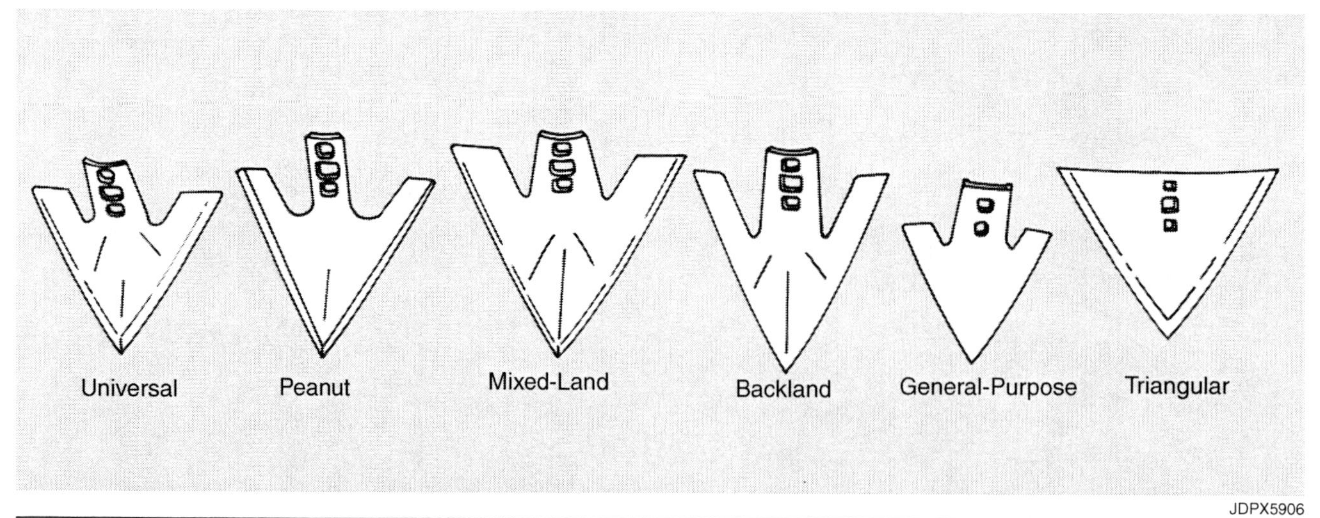


*Fig. 13 — Typical Range of Cultivator Sweeps*

## Summary

Weed control and breaking soil surface crust are major purposes of post-emergence tillage. Row-crop cultivator sweeps and shovels, and rotary cultivator and rotary hoe wheels slice off weeds, pull small weeds, and shatter the soil surface.

In addition to clean tillage, some row-crop cultivators are designed for use in reduced tillage, no-till, or other systems with heavy residue conditions.

Rotary hoes provide fast, economical cultivation before crop emergence or after plants are at least 2 inches (50 mm) tall. Rotary cultivators are similar in appearance to rotary hoes, but have wheels that turn backward and curved, slicing teeth.

## Test Yourself

### Questions

1. What are the two major objectives of post-emergence tillage?
2. (T/F) Row-crop cultivators are designed only for clean-till systems.
3. What are the purposes of rolling and hooded shields? Of spray shields?
4. What is the advantage of a spring-trip shank for row-crop cultivators?
5. Which tillage tool provides the fastest, most economical cultivation?
6. What speeds provide optimum performance of the rotary hoe?
7. (T/F) Rotary hoes operated at high speed often kill so many crop plants that yields are reduced.
8. How does the rotary hoe differ from the rotary cultivator?
9. (Fill in blank.) Front-mounted cultivators require more continuous steering attention than rear-mounted models because ______________________________.
10. (T/F) As plants mature, they become resistant to flame cultivation.

# Appendix

## SUGGESED READINGS

### TEXTS AND BULLETINS

*Agricultural Mechanics: Fundamentals and Applications*; Herren, Dr. Ray V.; Fourth Edition; Thomson Delmar Learning, Florence, Kentucky, 2006.

*The American Farm Book: A Practical Treatise on Every Staple Product of the United States, with the Best Methods of Planting, Cultivating, and Preparation for Market*; Allen, R. L.; The Lyons Press, 2002.

*CIGR Handbook of Agricultural Engineering ñ Volume I: Land and Water Engineering, Volume III: Plant Production Engineering, Volume VI: Information Technology*; American Society of Agricultural and Biological Engineers, St. Joseph Michigan, 1999.

*Conservation Tillage and Cropping Innovation: Constructing the New Culture of Agriculture*; Coughenour, C. Milton and Chamala Shankariah; Blackwell Publishing, Ltd., Williston, Vermont, 2000.

*Cultivating Utopia: Organic Farmers in a Conventional Landscape*; Hetherington, Kregg; Fernwood Publishing Co., Ltd., Nova Scotia, Canada, 2006.

*Farm Machinery: Practical Hints for Handy-men*; Davidson, J. Brownlee and Chase, Leon Wilson; The Lyons Press, 1999.

*Farm Power and Machinery Management*; Tenth Edition; Hunt, Donnell R.; Blackwell Publishing Professional, Williston, Vermont, 2001.

*Fundamentals of Machine Operation series – Tractors, Planting, Hay and Forage* Harvesting, Combine Harvesting, Preventive Maintenance, Machinery

*Maintenance*; John Deere Publishing, Dept. 373, 5440 Corporate Park Drive, Davenport, Iowa 52807.

*Managing Crop Residue with Farm Machinery*; Eck, Kenneth J., Brown, Darrell E. and Brown, Andrew B.; Agronomy Guide, Purdue University Cooperative Extension Service, AY-280, February, 2001.

*The Nature and Properties of Soils*; Thirteenth Edition; Brady, Nyle C. and Weil, Ray R.; Prentice Hall, Upper Saddle River, New Jersey, 2001.

*No-Tillage Seeding: Science and Practice*; Baker, C.J., Saxton, K.E. and Ritchie, W.R.; CABI Publishing, New York, 1996.

*Relationship Between Wheel-Traffic-Induced Soil Compaction, Nutrient Availability, and Crop Growth: A Review*; Wolkowski, R.P.; Journal of Production Agriculture, Vol. 3, No. 4, October-December, 1990.

*Safe Operations of Agricultural Equipment*; Hull, Dale and Silletto, Thomas; Third Edition; Hobar Publications, Lakeville, Minnesota, 1996.

*Soil Compaction in Crop Production*; First Edition; Elsevier Science, St. Louis, Missouri, 2000.

*Soils and Their Environment*; Hassett, John J. and Banwart, Wayne L.; Prentice Hall, Englewood, California, 1997.

*Soil Erosion and Conservation*; Third Edition; Blackwell Publishing Ltd., Williston, Vermont, 2005.

*Tillage Equipment: Lots of Room Left For Problem Solvers*; Buckingham, Frank; Implement & Tractor, Vol 101 (24), October 1, 1986.

*Trafficability Assessment for Agricultural Lands with the Aid of a Soil Compaction Model*; Marx, Barend, Bezuidenhout, Carel, Hansen, Alan C., Wilcox, Timothy A. and Lyne, Peter W. L.; American Society of Agricultural and Biological Engineers, Paper Number 061090, St. Joseph Michigan, 2006.

### DVDS AND VIDEOTAPES

*Agricultural Equipment Operator Safety Series*; 64-minutes; VHS seven-program series; (800) 423-5491, 1990.

*Agriculture Beyond 2000*; 30-minute documentary; American Farm Bureau Foundation for Agriculture, (202) 406-3701.

*Modern Marvels - Farm Technology*; 50-minute DVD; The History Channel, (888) 423-1212.

*Modern Marvels - Harvesting*; 50-minute DVD; The History Channel, (888) 423-1212.

### COMPUTER INTERACTIVES

*John Deere North American Farmer*; Farm simulation game for Windows ME/2000/XP.

## INSTRUCTOR'S GUIDE

FMO 11502T, teaching tips, class activities, quiz answers, masters based on FMO Tillage text. John Deere Service Publications, Dept. FOS/FMO, John Deere Road, Moline, Illinois 61265.

## MANUFACTURER'S LITERATURE

Performance-improving suggestions and safety tips are contained in the operator's manual supplied with each tractor and implement. If the manual for a particular piece of equipment is missing, contact the manufacturer for a replacement giving the make, model, year of manufacture (if known) and serial number of the unit.

## USEFUL INFORMATION

The following tables and charts are designed to serve as a quick reference to useful information related to tillage.

# Abbreviations

ANSI — American National Standards Institute

ASAE — American Society of Agricultural Engineers (sets standards for many hydraulic components for agricultural use)

bar — Metric unit of measure for pressure

C — Celsius (temperature)

F — Fahrenheit (temperature)

gpm — gallons per minute (fluid flow)

hp — horsepower

I.D — inside diameter

ISO — International Organization for Standardization (establishes many standards for worldwide use)

kg/cm$^2$ — kilograms per square centimeter (metric unit for pressure)

kPa — kilopascal (metric unit of measure for pressure)

kW — kilowatt (metric unit of measure for power)

lb-ft — pounds-foot (torque or turning effort)

lb-in. — pounds-inch (torque or turning effort)

L/m — liters per minute

N•m — newton-meter (metric unit of measure for torque)

O.D. — outside diameter

psi — pounds per square inch (pressure)

rpm — revolutions per minute

SAE — Society of Automotive Engineers (sets standards for many hydraulic components)

# MEASUREMENT CONVERSION CHART

| METRIC TO ENGLISH | ENGLISH TO METRIC |
|---|---|
| **Length** | **Length** |
| 1 millimeter = 0.03937 inches (in) | 1 inch = 25.4 millimeters (mm) |
| 1 meter = 3.281 feet (ft) | 1 foot = 0.3048 meters (m) |
| 1 kilometer = 0.621 miles (mi) | 1 mile = 1.608 kilometers (km) |
| **Area** | **Area** |
| 1 meter$^2$ = 10.76 square feet (sq. ft) | 1 square foot = 0.0929 meter$^2$ (m$^2$) |
| 1 hectare = 2.471 acres (acre)<br>(1 hectare = 10 000 m$^2$) | 1 acre = 0.4047 hectare (ha)<br>(1 hectare = 10 000 m$^2$) |
| **Mass (Weight)** | **Mass (Weight)** |
| 1 kilogram = 2.205 pounds (lb) | 1 pound = 0.4535 kilograms (kg) |
| 1 tonne (1000 kg) = 1.102 ton (tn) | 1 ton (2000 lb) = 0.9071 tonnes (t) |
| **Volume** | **Volume** |
| 1 meter$^3$ = 35.31 cubic feet (cu ft) | 1 cubic foot = 0.02832 meter$^3$ (m$^3$) |
| 1 meter$^3$ = 1.308 cubic yards (cu yd) | 1 cubic yard = 0.7646 meter$^3$ (m$^3$) |
| 1 meter$^3$ = 28.38 bushel (bu) | 1 bushel = 0.03524 meter$^3$ (m$^3$) |
| 1 liter = 0.02838 bushel (bu) | 1 bushel = 35.24 liter (L) |
| 1 liter = 1.057 quart (qt) | 1 quart = 0.9464 liter (L) |
| | 1 gallon = 3.785 liters (L) |
| **Pressure** | **Pressure** |
| 1 kilopascal = 0.145 pounds per square inch (psi)<br>(1 bar = 101.325 kilopascals) | 1 psi = 6.895 kilopascals (kPa)<br>1 psi = 0.06895 bars (bar) |
| **Stress** | **Stress** |
| 1 megapascal =145 pounds per square inch (psi)<br>(1 megapascal = 1 newton/millimeter$^2$) | 1 psi = 0.006895 megapascal (MPa)<br>1 psi = 0.006895 newton/millimeter$^2$ (N/mm$^2$) |
| **Power** | **Power** |
| 1 kilowatt = 1.341 horsepower (hp)<br>(1 watt = 1 Nm/s) | 1 horsepower (550 ft-lb/sec) = 0.7457 kilowatt (kW)<br>(1 watt = 1 Nm/sec) |
| **Energy (Work)** | **Energy (Work)** |
| 1 joule + 0.0009478 British thermal units (Btu) | 1 British thermal unit = 1055 joules (J) |
| **Force** | **Force** |
| 1 newton = 0.2248 pounds force (lb-force) | 1 pound = 4.448 newtons (N) |
| **Torque or Bending Moment** | **Torque or Bending Moment** |
| 1 newton-meter = 0.7376 pound-foot (lb-ft) | 1 pound-foot = 1.356 newton-meters (Nm) |
| **Temperature** | **Temperature** |
| °F = °C x 1.8 + 32 | °C = (°F – 32) / 1.8 |

# ANSWERS TO TEST YOURSELF QUESTIONS

## ANSWERS TO CHAPTER 1 QUESTIONS

1. Those mechanical, soil-stirring actions carried on for the purpose of nurturing crops.
2. more than one-half
3. six contributions of tillage to crop production:
   a. Management of crop residue
   b. Soil aeration
   c. Weed control
   d. Incorporation of fertilizer
   e. Moisture management
   f. Insect control
   g. Temperature control for seed germination
   h. Improvement of soil tilth
   i. Provide good seed-soil contact
   j. Prepare surface for other operations
   k. Erosion control
4. Primary; Secondary
5. False. Different implements and operations are required for different crops and soil conditions.
6. Damage to soil structure and soil compaction
7. Classifications of conservation tillage systems:
   a. No till
   b. Ridge-plant
   c. Strip till
   d. Disk and/or field cultivate
   e. Blade plow or chisel plow
   f. Subsoil
8. False. Reduced tillage systems do not always provide higher crop yields.
9. Steps in developing a plan for conservation farming:
   a. Evaluate existing situation
   b. Establish overall performance goals
   c. Identify specific improvements
   d. Identify alternatives
   e. Select conservation practices
   f. Evaluate progress toward goals
   g. Restate or modify goals if necessary
   h. Repeat steps as necessary
10. False. Conservation tillage systems do not always involve defined practices.
11. Provide a suitable environment for seed germination, root growth, weed control, soil-erosion control, and moisture control-avoiding moisture excesses and reducing stress of moisture shortages.
12. Growth in mechanization and changes in equipment designs and ideas.
13. False. Sales of moldboard plows fell 25 percent from 1967 to 1990.
14. The minimum soil manipulation necessary for crop production or for meeting tillage requirements under existing conditions.
15. Potential benefits and detrimental effects:
    a. Saves fuel and labor costs
    b. Permits earlier planting
    c. Reduce soil compaction
    d. Increased pesticide & fertilizer costs
    e. Possible lower yields
    f. Wet conditions can delay field operations
    g. Germination can be reduced

## ANSWERS TO CHAPTER 2 QUESTIONS

1. True. Soil compaction is caused primarily by surface pressure from wheel traffic.
2. Process of rearranging soil particles to decrease pore space and increase bulk density.
3. The arrangement of soil solids into aggregates; The proportion of sand, silt and clay in a soil.
4. Sand, Silt, Clay
5. False. Soil compaction is not always detrimental to crop growth.
6. Factors that affect crop response to soil compaction:
   a. Amount of compaction
   b. Crop species and variety

c. Growing season weather

d. Fertilizer management

7. 80-90%
8. 10 tons
9. False. Tracked vehicles cause more soil compaction than wheeled vehicles.

## ANSWERS TO CHAPTER 3 QUESTIONS

1. What is done with or what happens to the portion of plants remaining on the soil surface after the crop is harvested.
2. Above surface residues; below surface residues
3. Methods of measuring or estimating percent residue cover:

   a. Line-transect

   b. Photo comparison

   c. Prediction
4. False. The 1985 Food Security Act requires that 30 percent of the soil surface must be covered with residue in a conservation tillage system.
5. Composition and size of leaves and stems; Density and quantity of the residue.

   a. Examples of Fragile: Soybeans, Cotton, Peanuts, Sunflowers

   b. Examples of Non-Fragile: Corn, Alfalfa, Tobacco, Wheat
6. True. V-rippers/subsoilers leave more non-fragile residue on the soil surface than any other commonly used tillage implement.
7. Reduce soil erosion; increase soil organic matter content
8. A tillage system that leaves the maximum amount of crop residue on the soil surface.
9. Ways by which surface residues reduce soil erosion:

   a. Absorb the impact energy of the raindrops reducing splash erosion

   b. Provide an obstruction to flowing water slowing velocity of the runoff

   c. Reduce tendency for soil surface to seal which reduces infiltration
10. Moldboard plowing in the fall.

## ANSWERS TO CHAPTER 4 QUESTIONS

1. Proper tractor set-up factors:

   a. Total tractor weight and static weight split (weight distribution on front and rear axles)

   b. Type of ballast used (cast and/or liquid weight)

   c. Tire inflation pressures
2. The linear force, pull or draft, resulting from torque applied to tractor tires; the ability of tires to stay on top of the soil surface or resist sinking into the soil.
3. Benefits from a properly set-up tractor:

   a. Reduced compaction

   b. Improved ride

   c. Reduced tire wear

   d. Improved sidehill ability

   e. Better control of power hop on mechanical front wheel drive (MFWD) and four wheel drive (4WD) tractors
4. 51-55% on front axle and 49-45% on the rear axle.
5. True. Pulling a lighter load at higher speed can reduce wheel slip without increasing soil compaction.
6. Factors to monitor when tractor is at loads close to a traction or power limit:

   a. Wheel slip

   b. Engine speed

   c. Ground speed
7. 8-12% wheel slip
8. Improper adjustments or faulty components
9. The implement would operate opposite of the control which could be a hazard.
10. Before each season
11. False. A large, heavy duty bolt cannot be a satisfactory hitch pin.
12. Operator

## ANSWERS TO CHAPTER 5 QUESTIONS

1. They have not undergone the extensive testing of other machines, and it is difficult to predict the exact size of the bar, correct number of clamps, and lift capacity required.
2. By the use of square hallow bars.
3. Increased strength and additional rigidity
4. Toolbar attachment - Category, Specific operation, Crop:

   a. Sled carrier; Bed shaping; Beans, vegetables

   b. Spacer clamps; Precision row spacing; Cotton

   c. Gauge wheels; Precision depth of planting; Corn

   d. Standards and shanks; Subsoiler; Corn

   e. Soil-engaging tools; Row-crop cultivation; Soybeans

## ANSWERS TO CHAPTER 6 QUESTIONS

1. A more aggressive tillage action, working the soil at least six inches deep, leaves the surface rough, and completely inverts the soil to bury residue.
2. Soil is completely inverted burying surface residue; Moldboard plow, disk plow
3. Advantages of moldboard plow:
   a. Buries some or all trash and crop residue
   b. Aerates the soil
   c. Controls weeds, insects, and crop diseases
   d. Incorporate fertilizer into soil
   e. Provides good seedbeds for better germination

   Disadvantage of plow:
   a. Can form sole or hardpan and high horsepower requirements.
4. Key parts of plow bottom:
   a. Moldboard
   b. Share
   c. Shin
   d. Landslide
5. Tractor hydraulic-lift capacity and front-end stability.
6. Primary types of plowing disks:
   a. Single-action
   b. Double-action
   c. Tandem or double offset
   d. Offset
7. Levels and smooths the soil surface.
8. False. Use decreased with the advent of conversation tillage.
9. False. They perform best in soil that is dry and firm.
10. Because most crop residue remains on the surface after chisel plowing.
11. Types of soil engaging tools on chisel plows:
    a. Wheatland sweep
    b. High crown sweeps
    c. Heavy-duty low-crown sweeps
    d. Chisel sweeps
    e. Heavy-duty regular sweeps
    f. Furrow opener
    g. Beavertail shovels
    h. Reversible chisel points
    i. Double-point chisels
    j. Spikes
    k. Reversible double-point shovels
    l. Twisted shovels
12. Primary; Secondary tillage implements
13. By independently adjusting each part of the implement.
14. Chisel shanks
15. False. Mulch tillers are especially effective in heavy residue like cornstalks, stubble mulch or small grain fields.

## ANSWERS TO CHAPTER 7 QUESTIONS

1. Purposes of secondary tillage:
   a. Prepare fields for planting
   b. Summer fallowing
   c. Chemical incorporation
   d. Weed control
2. Seedbed preparation; Weed control; Stubble-mulch tillage; Roughening fields for moisture absorption and control of wind and water erosion
3. Chisel plows
4. Advantages of spring-cushion shanks for field cultivators:
   a. Permits points to lift over obstructions
   b. Provides shank vibration action for better soil pulverization
5. Disk; Coulter gang; Field cultivator
6. The amount of residue retained on the soil surface can be varied by choice of implements and soil engaging elements.
7. Type of soil engaging tool and depth of tillage
8. Types of toothed harrows:
   a. Fine-tooth
   b. Spike-tooth
   c. Spring-tooth
9. Main uses of toothed harrows:
   a. Smooth seedbeds
   b. Break soft clods
   c. Kill small weeds
10. A heavy-duty wheeled spring-tooth harrow, built wider and stronger.
11. Cultipackers; Soil pulverizers; Corrugated rollers

12. Crush clods and firm the soil surface, break soil crust, close air pockets
13. Basic differences between roller harrows and tooth-type harrows:
    a. Roller harrows crush clods and level the surface.
    b. Tooth harrows break soil crust, shatter clods, kill weeds, loosen and aerate soil, and cultivate small plants.

## ANSWERS TO CHAPTER 8 QUESTIONS

1. 3-6 inches
2. Hot
3. Anchoring residue firmly in place; Prepare soil surface for planting.
4. True.
5. Bedders; Lister-planters
6. Profile planting:
    - Putting the crops on a small ridge or raised bed in the bottom of the lister furrow.

    Advantages:

    - Places seed in moist soil, protects young plants from wind and blowing soil, reduces the chances of water standing on the row or washing over the seed or seedlings during a heavy rain.
7. In high-rainfall areas
8. False. Bedders are normally classified as primary tillage tools.
9. Where lister bottoms do not scour or there are many soil obstructions.
10. Front side rotates up to pull up and get rid of roots.
11. Western wheat growing areas. For summer fallow and just before seeding.

## ANSWERS TO CHAPTER 9 QUESTIONS

1. Types of tillage attachments for row-crop planters:
    a. Disk furrower
    b. V-wing attachment
    c. Tine-tooth tillage attachment
    d. Coulters
    e. Row cleaner attachment
2. The fluted coulter prepares a seedbed 3 inches wide while the ripple colter prepares a seedbed about 0.75 to 1.5 inches wide.
3. Functions of press wheels on press-wheel drills:
    a. Firm the soil over the seed
    b. Drive the seed metering mechanism
    c. Support the rear of the drill
4. Applications for tilling seeder:
    a. Plant new pastures
    b. Renovate old pastures
    c. Seed roadsides, parks and golf courses
5. Seed tank; Tillage tool; Chisel plow; Field cultivator; Sweep blades
6. True.
7. Air seeders deliver the seed by air while seed is delivered by gravity with grain drills.
8. False. Seed equipment is successful in full-tillage and conservation tillage systems.
9. Small grains; Soybeans
10. False. It can be towed behind or in front of the seeding tool.

## ANSWERS TO CHAPTER 10 QUESTIONS

1. Major objectives of post-emergence tillage:
    a. Weed control
    b. Preparation of land for irrigation
    c. Breaking soil surface crust
2. False. Row-crop cultivators, specifically some models of rear-mounted cultivators, are designed for use in reduced tillage systems or other heavy residue conditions.
3. Rolling and hooded shields - to prevent covering plants with soil.

    Spray shields - hold spray in confined area
4. To cause the shank to trip and automatically reset when an obstacle is encountered.
5. Rotary hoe
6. 7–12 mph (11–19 km/h)
7. False. While extremely high speeds may tear leaves of crop plants and cause some uprooting, the benefits of rotary hoeing more than offset any crop mortality or crop yield.
8. Rotary cultivator wheels turn backwards, and have curved slicing teeth, teeth are also designed for left and right hand cutting.
9. They do not have the lateral flexibility of the rear-mounted equipment to average out steering corrections or deviations.
10. True.

# TILLAGE HISTORY

## SOME HISTORIC EVENTS:

**6000 B.C.** – Egyptian drawings show a forked stick, apparently with a stone point, being used as a hoe or mattock. Later one fork was left longer for pulling by slaves or animals.

**900 B.C.** – Elisha was found plowing with twelve yoke of oxen before him. (Kings 19:19)

The plow changed little in the next 26 centuries until the 18th century A.D. In 1721 the Norfolk wheel-plow had a cast-iron share and rounded iron moldboard.

**1730** – The Roman plow was brought to northern Europe.

**1750's** – The Essex plow had an iron moldboard.

**1760** – The curved moldboard appeared on Suffolk swing plow.

**1797** – Charles Newbold obtained the first United States patent for a plow. The idea of a one-piece cast-iron share, moldboard, and landside was rejected because of high replacement cost. Farmers thought iron plows poisoned the soil and caused weeds to grow.

**1798** – Thomas Jefferson designed a moldboard plow from mathematical computations. He hoped to design an ideal shape for all soil's.

**1813** – R. B. Chenoworth of Baltimore patented a cast-iron plow with separate share, moldboard, and landside.

**1813, 1819** – Jethro Wood received patents on cast-iron plows. He developed a curved moldboard to turn soil in even furrows.

**1820** – The horse hoe with one large shovel was developed for row-crop cultivation. A second shovel was added to the design about 1850,

**1833**--John Lane made the first steel plow from three sections of an old handsaw.

**1837**-John Deere used an old sawmill blade to shape a one-piece steel plowshare arid moldboard over a log pattern (Chapter 1, Fig. 1). His highly polished steel plow turned the sticky prairie soil where cast-iron plows failed.

**1846** – The first one-horse wheel cultivator was used. **1847** – The disk plow was patented.

**1856** – M. Farley patented a single-bottom, wheeled riding sulky plow.

**1856** – A 2-horse, straddle-row walking cultivator was patented; it cultivated both sides of the row simultaneously.

**1860's** – The Civil War, with shortage of labor and booming demand for food and fiber, spurred improvements in all types of farm equipment.

**1860's** – Chisel cultivators came into use.

**1863** – Riding cultivators became a commercial success.

**1864** – F. S. Davenport patented a 2-bottom horse-drawn gang plow.

**1868** – John Lane, who made a steel plow from sections of hand saw, patented a soft-center steel moldboard (currently used in making most plow moldboards).

**1869** – The springtooth harrow was patented. 1877-The disk with concave blades was patented.

**1880** – Keystone Manufacturing Co. started factory production of disk harrows in Sterling, IL.

**1880** – Commercial lister production was underway. 1884-The first 3-wheel riding plow was used. 1890-The disk plow was developed for practical use.

**1900**'s – The 2-row horsedrawn cultivator came into use.

**1910** – The rod weeder was developed. 1912-The rotary hoe was produced commercially.

**1914-1918** – World War I labor shortage and demand for agricultural products accelerated farm-mechanization.

**1918** – B. F. Avery Co. built a tractor-mounted row-crop cultivator.

**1920's** – Mechanical power lifts were developed for moldboard plows.

**1924** – The offset disk was developed.

**1927** – The disk tiller, or one-way wheatland disk plow, began to sell in large numbers.

**1930** – Swiss-made rotary tillers were introduced in the United States.

**1930**'s – Power lifts for cultivators were developed. Harry Ferguson developed the integral plow and tractor 3-point hitch (in England). These were brought to the United States in 1939.

**1935** – The National Tillage Machinery Laboratory was established by the United States Department of Agriculture at Auburn, AL

**1941** – Hydraulic remote control of drawn implements was introduced.

**1941-1945** – World War II intensified demand for mechanized agricultural production of more crops with less labor.

**1949** – The wheel disk was introduced. 1953-Safety-trip beam plows were in use.

**1960**'s – Rotary cultivators and selective crop thinners were introduced.

**1960**'s – The hydraulic reset plow was introduced.

**1970**'s – Combination disk/chisels (stubble-mulch tiller) and disk/field cultivators were introduced and perfected.

**1980**'s – On board computers and electronic sensors increase efficiency.

**1980**'s – Combination secondary tillage tools such as disk/field cultivator were introduced.

**1992** – Development of secondary tillage tools designed to retain rather than destroy surface residue.

Parallel development of other equipment and practices or processes have increased the pressure for development and acceptance of many new tillage methods and machines.

To name a few (not in chronological order):

1. Invention of the steam engine which helped spark the industrial revolution.

2. Development of steam, gasoline and diesel tractors.

3. Invention of cotton gin.

4. Invention of reaper, cornpicker, and grain combine.

5. Adaptation of rubber tires to farm tractors.

6. Invention of mechanical planters and drills.

7. Development of balers, forage harvesters, and other hay tools.

8. Beginning of adoption of no-till systems in late 1960's.

9. Increased fuel prices after 1973.

10. Adoption of conservation tillage systems in late 1980's.

# GLOSSARY

AUTOMATIC RESET STANDARDS – Mechanism for protecting soil-working tools by swinging backward and up if solid soil obstruction is encountered, then immediately resuming operating position when obstacle is passed. May be hydraulic or mechanical.

BALLAST – Weight added to tractor or implement to improve tractor traction and stability and/or improve penetration of soil-working tool.

BEDDER – Primary tillage implement which makes wide V-shaped furrows in or on which row crops are planted.

BED SHAPER – Implement that makes wide beds on which rows of crops are planted. Furrows between beds are used for irrigation and to guide cultivators.

CENTER OF LOAD – Vertical and lateral center of resistance on tillage implements; affected by soil conditions, depth of tillage, type of soil-working tools, and operating speed.

CENTER OF PULL – Vertical and lateral center of pull of tractor; affected by tractor rear-wheel setting, convergence of 3-point hitch links with integral implements, and front pivot point of tractor drawbar with drawn implements.

CHISEL PLOW – Primary tillage implement which breaks and shatters soil to leave a rough residue-covered surface. Basic frame has rows of staggered, curved shanks for various types of soil working tools, including points, sweeps, chisels, spikes and shovels.

CHISELS – Deep-working, narrow soil-engaging tools for chisel plows.

CLEARANCE – Vertical distance between implement frame bars and soil surface or lateral distance between standards for soil-working tools.

CLEAN TILLAGE – Sequence of operations which prepares a seedbed having essentially no plant residues on the soil surface.

CLOD BUSTER – Tooth-type harrow attachment for rear of primary-tillage implements.

COMBINATION IMPLEMENT – A single or group of tillage tools operating as a single machine.

COMPACTION – See Soil Compaction.

CONE GUIDE – Used on cultivators for precision tillage of crops on bed-shaped land.

CONE-SHAPED BLADE – Concave disk blade which has equal distance between working surfaces of adjacent blades, top to bottom, for easier soil movement and less soil compaction.

CONSERVATION TILLAGE – Leaving some or much residue on soil surface when preparing soil, planting, and cultivating. Objectives are reduced soil erosion, moisture retention in soil, reduced compaction, and saving of fuel, time, and labor.

CONTROLLED TRAFFIC – System in which all wheel traffic from tillage through harvest is confined to same tracks across field, with that area left untilled.

CONVENTIONAL TILLAGE – Sequence of tillage operations traditionally or most commonly used in a specific geographical area.

CROP RESIDUE – Organic matter from previous crop left on soil surface after harvest. Includes stalks, stubble and weeds. Also called trash.

CROP SHIELDS – Operate close to or over crop row (1) during cultivation to keep soil from covering small crop plants, or (2) during herbicide application to keep chemical from striking crop plants.

CORRUGATED ROLLER – See Roller Packer.

COULTERS – Sharp steel disks, usually flat, used to cut trash and define furrow slice ahead of moldboard plow bottoms, leaving clean furrow wall and reducing wear on share and shin; to provide lateral stability for rear-mounted row-crop cultivators and disk tillers; or assembled in rows and attached to the front of chisel plows (stubble mulch tillers) or field cultivators to cut through residue and reduce plugging. May have plain, notched or ripple-edge blades, and some are concave rather than flat. See Disk Coulter.

CULTIPACKER – See Roller Packer.

DEEP-TILLAGE BOTTOM – Plow bottom with extra-high moldboard for deep work in heavy soil.

DISK BEDDED – See Bedder.

DISK (DISK HARROW) – Tillage implement which moves soil both right and left by means of concave spherical or conical blades.

DISK HILLER – Cultivating tool with small concave blades instead of sweeps, shovels, or knives.

DISK PLOW – Primary tillage implement cuts, lifts and rolls furrow slice. Differs from other disk implements in that each concave blade has its own bearing and standard and revolves independently of the others.

DISKS – Concave, spherical or conical rotating blades for uses ranging from shallow row-crop cultivation to secondary tillage to deep primary tillage.

DISK TILLER – Intermediate between disk plow and disk, this tool is for both, primary and secondary tillage. Uses concave spherical disk which turn together on common shaft. Also known as one-way, seeding tiller, wheatland disk plow, and wide-level disk plow.

DO-ALL – See Finishing Harrow.

EFFECTIVE FIELD CAPACITY – Actual work accomplished (acres or hectares per hour) by an implement despite loss of time from field-end turns, inadequate tractor capacity, deficient tractor or implement preparation, adverse soil conditions, irregular field contours, lack of operator skill, or other factors. See theoretical field capacity, field efficiency.

FIELD CAPACITY – The maximum amount of water a soil can hold after free drainage.

FIELD CONDITIONER – Essentially heavy-duty wheeled spring-tooth harrow for secondary tillage; uses variety of soil-working tools.

FIELD CULTIVATOR – Similar to chisel plow but lighter in construction, designed for less-severe conditions, and with less vertical and lateral trash clearance. Has points, shovels, or sweeps on flexible shanks.

FIELD EFFICIENCY – Ratio of the area an implement can theoretically cover in one hour and the area actually covered. Ratio is expressed as a percentage of theoretical field capacity.

FIELD LAYOUT – Planning work to reduce number of row-end turns, point rows, and back and dead furrows.

FINISHING HARROW – Variation of rolling packer consisting of a pair of 5-bladed, horizontal cutterheads to chop and crush clods and firm the soil. Final finishing and smoothing is accomplished by a leveling plank mounted across rear of machine. Also called do-all.

FLAME CULTIVATION – Flares of burning LP gas kill young weeds between and sometimes (if crop plants are tall enough) within crop rows.

FLEXIBLE-TINE CULTIVATOR – See S-tine.

FLOTATION – Ability of tractor or implement tires to stay on top of soil surface; usually related to soil condition, tractor or implement weight, wheel slippage and contact area between tires and soil surfaces.

FRAGILE RESIDUE – Crop residue that is essentially all damaged in passing through the combine. The stems are small in diameter, the leaves are small in size and leaves tend to fall from the plant before harvest. May also have high nitrogen content.

FROG – Part of plow bottom which connects moldboard, share, shin., and landside.

FURROW FILLERS – Smaller disk blades at outer ends of rear gangs of disk harrows.

FURROW WHEEL – Tractor or implement wheel which runs in furrow from previous implement pass.

GAUGE WHEELS – Control working depth and improve stability of many tillage implements; sometimes double as transport wheels.

GENERAL-PURPOSE BOTTOM – Plow bottom with long, fairly slow-turning (of furrow slice) moldboard.

GUIDE FINS – Used on some row-crop cultivators for lateral stability.

HIGH-SPEED BOTTOM – Plow bottom with longer moldboard than general-purpose bottom and less curve at upper end.

IN-FURROW HITCH – One tractor rear wheel runs in previous furrow when plowing.

INVERSION PRIMARY TILLAGE – Primary tillage which partially or completely inverts the soil to bury residue from the previous crop.

JOINTERS – Cut small ribbons of soil from surface just ahead of plow share points for improved trash coverage.

KNIVES – Shallow-working tools for killing weeds close to such crops as beets, beans, and vegetables with minimum surface disturbance.

LANDSIDE – Part of plow bottom which bears against the furrow wall to offset side draft, improve plow stability, and sometimes carry part of weight of plow; may be flat or rolling.

LIFT-ASSIST WHEELS – For integral implements which tax or exceed hydraulic lift and/or front-end stability limits of tractors.

LINE OF DRAFT – Theoretical line between tractor center of pull and implement center of resistance.

LISTER – See Bedder.

LISTER-PLANTER – Lister with planting attachment for once-over soil preparation and planting.

MIDDLEBREAKER – See Lister; also name for attachment to remove ridge of soil between opposed gangs of disks.

MINIMUM TILLAGE – System of conservation and mulch tillage using minimum number of operations necessary to produce an acceptable yield. May involve combined or reduced operations. Also called optimum, reduced and economy tillage.

MOLDBOARD – Part of plow bottom which fractures, crumbles, and inverts furrow slice.

MOLDBOARD-EXTENSION – Used at rear of plow moldboard for more-positive control of furrow slice on hillsides or in heavy trash.

MOLDBOARD PLOW – Primary-tillage implement which severs bottom and one side of furrow slice, fractures and granulates the soil, inverts it into the previous furrow, and buries surface trash.

MULCH TILLAGE – See Conservation Tillage.

MULCH TILLERS – Hybrid implements which combine disk and chisel-plow soil-working tools and functions. Some models have flat coulter blades instead of concave disk blades.

NON-FRAGILE RESIDUE – Crop residue that is resistant to damage from the combine. Stems and leaves are larger than from fragile crop residues.

NON-INVERSION PRIMARY TILLAGE – Primary tillage that disturbs surface residue as little as possible.

NO-TILL DRILL – Heavy duty grain drill designed to plant seed into soil untilled after the previous crop.

NO-TILL PLANTING – Planting in soil undisturbed from previous crop except for tillage provided by planter or drill.

OFFSET DISK – Has two opposed disk gangs, working one behind the other.

ONE-WAY – See DiskTiller.

ON-LAND HITCH-All tractor wheels run on untilled soil.

OPTIMUM TILLAGE – See Minimum Tillage.

PLOW BOTTOM – Three-sided wedge with landside and share as flat sides and shin and moldboard as curved side which inverts furrow slice severed by share and shin.

PLOWING DISK – Heavy duty offset disk with large blades and ample strength for deep primary tillage.

PLOW PAN – See Plow Sole.

PLOW SOLE – Compacted layer, restricting root and water movement, which may form in some soils just below the tilled area after several years of primary tillage to the same depth.

POWER HARROW – Secondary tillage implement with reciprocating tines powered by tractor PTO.

POWER HOP – Simultaneous loss of tractor traction and a bouncing, pitching ride. Occurs most frequently under high drawbar loads in certain soil conditions when using mechanical front wheel or four-wheel drive tractors.

PRESS-WHEEL DRILL – Grain drill with press-wheel gangs mounted on the rear to firm soil over the seed, drive the seed metering mechanism and support the rear of the drill.

PRIMARY TILLAGE – Tillage that cuts and/or shatters soil and may bury residue by inversion, mix it into the tilled layer or leave it basically undisturbed. An aggressive operation that typically tills to a depth of 6 inches (15 cm) or deeper.

PROFILE PLANTING – Variation of lister planting in which crop is planted on small ridge or raised bed in bottom of lister furrow.

REDUCED TILLAGE – -Tillage system which uses fewer operations than conventional tillage for a particular crop or region.

RESIDUE – See Crop Residue.

REVERSIBLE BOTTOM – Single bottom on reversible plow that is hydraulically reversed at each end of the field.

REVERSIBLE PLOW – Moldboard plow designed with rollover (two sets of bottoms) or reversing (one set of bottoms) capabilities. Bottoms are alternated or reversed at each end of the field so that all furrows can be turned in the same direction. Also called two-way plow.

RIDGE PLANTING – System in which crops are planted and grown on pre-formed ridges. May be a form of no-till or clean-till system.

ROD WEEDER – Secondary-tillage implement used primarily for fallow-land weed control with minimum soil-surface disturbance and moisture loss. Employs round or square weeding rods, powered from gauge wheels, rotating under soil surface.

ROLLER HARROW – Field-finishing or pre-planting implement with one row of heavy rollers, two rows of spring teeth, then another row of rollers. Roller types include solid-rim, serrated, crowfoot, and sprocket-type; they operate independently on tubular gang axles.

ROLLER PACKER – Has rollers, similar to roller harrow, but no spring teeth.

ROOT CUTTER – Short horizontal blade attached to plow-bottom landside for more-positive cutting of heavy roots.

ROTARY CULTIVATOR – Tillage implement with rigid, curved teeth mounted on wheels much like rotary hoe, but teeth are angled to slice and twist for more aggressive weed control and tillage.

ROTARY HOE – Tillage implement that permits fast, shallow cultivation before, or soon after, crop plants emerge. Rigid, curved teeth roll over the ground, penetrate almost straight down, but lift soil as they emerge.

ROTARY TILLER – Tillage implement that uses C-shaped or L-shaped overlapping blades mounted on flanges and rotors. Powered from tractor PTO. Used for many functions ranging from row-crop cultivation to full depth soil conditioning for primary tillage.

ROW-CROP CULTIVATOR – Tillage implement with gangs or rigs of sweeps, knives, shovels, disk weeders, disk hillers or other soil working tools which run between crop rows to break crust, kill weeds, aerate soil and improve moisture absorption.

SAFETY TRIPS – Mechanical devices on plows and other tillage implements which permit standards or the soil-working tools to release backward if they strike solid obstructions.

SCRAPERS – Attachments for tillage disks and steel or cast-iron implement wheels which help prevent dirt and trash buildup and plugging.

SECONDARY TILLAGE – Tillage that is shallower than primary tillage, provides additional pulverizing, levels and firms the soil, kills weeds and helps conserve moisture.

SEEDING TILLER – See Disk Tiller.

SHANKS – Standards extending down from implement frames to which soil-working points, shovels, and sweeps of row-crop cultivators and other tillage implements are attached; may be rigid, flexible, or spring-cushioned.

SHARE – Part of plow which severs bottom of furrow slice.

SHEAR BOLTS – Protect soil-engaging tools, such as plow bottoms, by shearing and releasing standards if solid soil obstruction is encountered.

SHOVELS – Deep-working (compared to sweeps and knives) soil-working tools for row-crop and field cultivators, chisel plows, and other implements.

SINGLE-ACTION DISK – Has two opposed disk gangs working side by side.

SLATTED BOTTOM – Plow bottom with slatted moldboard for better scouring in sticky soils.

SLED CARRIER – Toolbar with flat steel sleds for precision planting and cultivating of crops on bed-shaped land.

SLED GUIDE – Used on cultivators for precision tillage of crops on bed-shaped land.

SLIPPAGE – See Wheel Slip.

SMOOTHING HARROW – Tooth-type harrow attachment for rear of disks and plows.

SMV EMBLEM – Used at rear of tractors and implements during road transport to warn overtaking car and truck drivers of slow-moving vehicle.

SOIL COMPACTION – Process of rearranging soil particles that decreases pore space and increases bulk density. Usually detrimental to crops, but may also be beneficial in promoting seedling emergence.

SOIL ORGANIC MATTER – Decomposed plant residue component of the soil mass; important in aggregation of soil particles (soil structure), moisture absorption, and fertility of soil.

SOIL PORE SPACE – Space between soil aggregates. Soil moisture and air occupy these pore spaces.

SOIL STRUCTURE – Arrangement of soil solids (sand, silt and clay particles) into aggregates bound together by organic matter.

SOIL TEXTURE – Designation of a soil by the proportion of sand, silt and clay in its aggregates. Examples: silty clay, silt loam, coarse sand.

SPIKES – Deep-working, narrow, soil-engaging tools for chisel plows; often used to rip hardpan or plow sole.

SPIKE-TOOTH HARROW – Secondary-tillage implement for seedbed preparation, crust breaking, killing small weeds, and similar purposes.

SPRING-TOOTH HARROW – Aggressive secondary-tillage implement for such operations as deep seedbed preparation and destroying persistent weeds. Has vibrant spring-steel shanks for variety of soil-working tools.

SQUADRON HITCH – Wide hitch for using two or more implements working end to end. See Tandem Hitch.

STANDARDS – Supports for soil-working tools of primary-tillage implements, such as bottoms on plows. Have various devices to prevent damage from solid soil obstacles, including shear bolts, safety trips, and hydraulic or mechanical automatic resets.

STATIC WEIGHT SPLIT – Relative weight distribution on front and rear axles of a tractor.

S-TINE (FLEXIBLE) CULTIVATOR – Cultivator with sweeps or shovels on vibrant, S-shaped or double-curved spring-steel shanks; usually works deeper than spike or tine-tooth harrows.

STUBBLE BOTTOM – Plow bottom with short, abruptly curved moldboard which turns furrow slice quickly; used where scouring is difficult; not suitable for fast speeds.

STUBBLE-MULCH PLOW – Plow with shallow-running sweeps or blades which cut weeds but leave residue anchored on soil surface with minimum soil disturbance and moisture loss.

STUBBLE TREADER – Angled gangs of curved teeth that kill weeds and anchor residue to reduce erosion in dryland areas. Used independently or attached to wide-sweep plow.

SUBSOILERS – Chisel points or other soil-working tools operated below normal tillage depth to break up impervious soil layers and improve root and water penetration.

SWEEPS – Soil-working tools with wide cutting edges for cultivators, chisel plows, stubble-mulch plows and other tillage implements.

TANDEM DISK-X – shaped disk with two opposed front gangs and two opposed rear gangs.

TANDEM HITCH – Two or more implements hitched one behind another. See Squadron Hitch.

THEORETICAL FIELD CAPACITY – Work which could be done by an implement at optimum speed if no time were lost. See Effective Field Capacity.

TILLAGE – Mechanical soil-stirring actions for nurturing crops by providing suitable soil environment for seed germination, root growth, and weed and moisture control.

TILLING SEEDER – Tillage implement that falls between disk and moldboard plow in function. Consists of spherical blades mounted on a common axial shaft that prepare a seedbed for each seed drop on the drill. Normally throws soil to the right. Considered mostly as a dryland implement.

TILL PLANTER – Planter that operates in seedbed with little or no preliminary tillage after the previous crop.

TINE-TOOTH HARROW – Round, flexible teeth extending down from several rows of frame bars. Less aggressive than spike-tooth harrow.

TOOLBARS AND TOOL CARRIERS – Basic frames for assembling do-it-yourself tillage and planting implements from scores of component parts, available from implement dealers, including gauge and transport wheels, standards and shanks, soil-engaging tools, hitches, braces, clamps, markers, hydraulic cylinders and hoses, unit planters, and others.

TRACTION – Effective force resulting from thrust of tractor tires against soil or other surface; depends on such factors as nature of surface, contact area between tires and surface, and tractor power and weight.

TRASH – See Crop Residue.

TRASHBOARDS – Used above plow moldboards for more-positive trash deflection into furrow bottom.

TREADER – See Stubble Treader.

TWISTED SHOVELS – Curved, somewhat like plow moldboard, to provide more soil inversion and residue coverage than flat shovels used on chisel plows.

TWO-WAY PLOW – See Reversible Plow.

WALKING BEAM – Improves surface conformity of tillage implements equipped with gauge wheels by having one wheel at each end of short support frame pivoted under main implement frame.

WEED HOOKS – Used on moldboard plows to hold tall weeds and bulky residue against furrow slice for improved covering.

WHEATLAND DISK PLOW – See DiskTiller.

WHEEL SLIP – Reduced forward speed of tractor drive wheels resulting from soil conditions that limit total traction. Excessive slippage is detrimental, but some slippage is desirable to cushion tractor engine and drive train from sudden overload.

WIDE-LEVEL DISK PLOW – See Disk Tiller.

WIDE-SWEEP PLOW – See Stubble-mulch Plow.

ZERO-TILL PLANTING – See No-till Planting.

# INDEX

## G



## H



## I



## J



## L



## M



## N



## O



## P

## V